SpringerBriefs in Applied Sciences and Technology

SpringerBriefs present concise summaries of cutting-edge research and practical applications across a wide spectrum of fields. Featuring compact volumes of 50–125 pages, the series covers a range of content from professional to academic.

Typical publications can be:

- A timely report of state-of-the art methods
- An introduction to or a manual for the application of mathematical or computer techniques
- A bridge between new research results, as published in journal articles
- A snapshot of a hot or emerging topic
- An in-depth case study
- A presentation of core concepts that students must understand in order to make independent contributions

SpringerBriefs are characterized by fast, global electronic dissemination, standard publishing contracts, standardized manuscript preparation and formatting guidelines, and expedited production schedules.

On the one hand, **SpringerBriefs in Applied Sciences and Technology** are devoted to the publication of fundamentals and applications within the different classical engineering disciplines as well as in interdisciplinary fields that recently emerged between these areas. On the other hand, as the boundary separating fundamental research and applied technology is more and more dissolving, this series is particularly open to trans-disciplinary topics between fundamental science and engineering.

Indexed by EI-Compendex, SCOPUS and Springerlink.

More information about this series at http://www.springer.com/series/8884

João M. P. Q. Delgado ·
Antonio Gilson Barbosa de Lima ·
Mariana Julie do Nascimento Santos

Transport Phenomena in Liquid Composite Molding Processes

João M. P. Q. Delgado
Faculty of Engineering, CONSTRUCT-LFC
University of Porto
Porto, Portugal

Mariana Julie do Nascimento Santos
Department of Mechanical Engineering
Federal University of Campina Grande (UFCG)
Campina Grande, Paraíba, Brazil

Antonio Gilson Barbosa de Lima
Department of Mechanical Engineering
Federal University of Campina Grande (UFCG)
Campina Grande, Paraíba, Brazil

ISSN 2191-530X ISSN 2191-5318 (electronic)
SpringerBriefs in Applied Sciences and Technology
ISBN 978-3-030-12715-2 ISBN 978-3-030-12716-9 (eBook)
https://doi.org/10.1007/978-3-030-12716-9

Library of Congress Control Number: 2019930633

This Springer imprint is published by the registered company Springer Nature Switzerland AG
The registered company address is: Gewerbestrasse 11, 6330 Cham, Switzerland

Preface

The progress in the field of composite materials has been given for many professionals from the chemical, materials, mechanical and manufacturing engineering, and other scientific areas.

In the present days, the development of new and advanced composite manufacturing processes has allowed the manufacture of innovative products with high quality from the technological and structural performance points of view. These products are being made by several industries such as automotive, marine, aerospace, sports, and products.

In the majority, they are manufactured by the liquid molding process, with special reference to resin transfer molding (RTM) process. In fact, this technique has gained the preference because it offers several advantages in terms of high surface finish quality and good mechanical properties to the manufactured products.

Despite the importance, unfortunately, this topic has been shortly discussed in textbooks available in the literature, which provides a gap for new academic books aimed at this issue. In this sense, this book is intended to provide valuable information about polymer composite manufacturing with emphasis in liquid molding processes and the resin transfer molding (RTM) technique. In this document, emerging topics related to foundations, engineering applications, advanced modeling, and experiments of the RTM process are presented and discussed.

For many years, we have the enormous pleasure of working with various talented researchers including our current and former students (undergraduate, graduate, and postgraduate levels), in the theme "composite manufacturing." Then, the intention of this book is to document our research progress at this field.

The book contains five chapters about composite materials manufacturing with particular reference to RTM process. Following the steps of the book, in Chap. 1, an introduction is given to the polymer composite reinforced with fiber, and the motivation to study the liquid composite molding process. Chapter 2 is devoted specially to the resin transfer molding technique. The main issues, including foundations, processing stages, process control, advantages and disadvantages associated with the use of the RTM technique and application at different areas, are presented and discussed. The experimental techniques that have been developed by our research group related to resin transfer molding are presented in Chap. 3. Two important topics related to the RTM process are presented: rectilinear and radial infiltration processes, with or without fillers (reinforcements). Chapter 4 presents a comprehensive and rigorous analysis about fluid flow in porous media with particular emphasis to RTM process. In this chapter, different approaches (analytical and numerical) for description of air and resin flow in fibrous media are given. An advanced mathematical modeling is proposed which includes different effects of process parameters such as injection pressure, fluid viscosity, porous media permeability and porosity, and mold geometry. In Chap. 5, we present our conclusions about the different chapters' content and the main results present in them. The idea is to help professionals, engineers, industrials, and academics involved in this advanced and interdisciplinary field.

Porto, Portugal — João M. P. Q. Delgado
Campina Grande, Brazil — Antonio Gilson Barbosa de Lima
Campina Grande, Brazil — Mariana Julie do Nascimento Santos

Acknowledgements

We would like to thank Prof. Dr. Sandro Campos Amico, Jeferson Avila Souza, Brauner Gonçalves Coutinho, Iran Rodrigues de Oliveira, and Felipe Ferreira Luz for providing excellent researches about this field and contributions in this book.

We thank our Publisher and Editors for making valuable suggestions to the contents of the book, and especially to Prof. Dr.-Ing. Andreas Öchsner for believing in this project. Finally, we express our gratitude for each one of the cited authors and researchers, and also to the Portuguese and Brazilian (CNPq, CAPES, and FINEP) research agencies for the financial support.

Contents

Chapter 1
Introduction

Abstract This chapter provides important information about composite materials with emphasis to polymer composite reinforced with fibers, and the motivation to study the liquid composite molding process. Herein, different topics such as definition, classification, constituents, technological characteristics, manufacturing techniques, and performance of polymer composites are presented and discussed.

1.1 Motivation

Composite materials have received special attention for many years, because of the wide range of applications in different areas (boats, ships, automotive and aircraft components, submersibles, offshore structures, prostheses, etc.). Fibers reinforced polymer composite materials are preferred to designing structural materials when compared to conventional materials (metals and ceramics). The reasons are related to higher mechanical properties of strength and stiffness of the fibers.

However, the mechanical properties of these materials mainly depend on the adhesion between fiber and polymer, which are strongly dependent of manufacturing techniques. One of them is the liquid composite molding process, especially the resin transfer molding.

Basically, the RTM process consists in injecting the resin in a fibrous media located in a mold cavity until the fulfillment of the cavity. This physical situation is similar to fluid flow in porous media at a macroscopic scale.

Despite of the importance, few books offer detailed information about liquid composite molding process, including fluid flow mathematical models and its analytical and numerical solutions.

Mathematical modeling and numerical computation are processes of obtaining approaches for solutions of problems of academic and/or industrial interest. RTM physical problems involves partial differential equations of mass, momentum and energy conservations as applied for porous media. In these areas, several books dedicated to this theme, provide an understanding about resin flow through the mold cavity only under rectilinear and/or radial infiltration. Therefore, information addressed

J. M. P. Q. Delgado et al., *Transport Phenomena in Liquid Composite Molding Processes*, SpringerBriefs in Applied Sciences and Technology,
https://doi.org/10.1007/978-3-030-12716-9_1

to resin and air two-phase flow through mold with complex geometries is practically null. Then, there is a lack in this topic.

In this sense, this comprehensive book responds to the large growing interest in polymer composite molding processes, specially that involving mathematical treatment of the governing equations, experiments and numerical applications. It is the first book entirely dedicated to this subject.

The key challenge of this book is to document different information related to fiber-reinforced polymer composite ranging from basic material and manufacturing to advanced flow mathematical modeling in fibrous media and applications, in a unique volume.

The idea is to provide an in-depth analysis of the key issues including rigorous and coupled engineering models, experimental results, and general foundations about the polymer composites manufacturing process.

Special attention is given to resin transfer molding in different conventional and non-conventional geometries and for different two-phase infiltration processes (rectilinear, radial and arbitrary) including sorption and curing phenomena of the resin.

The consolidated information in this book will help professional, curious readers, undergraduate and graduate students, as well as applied mathematicians, engineers and scientists to better understand advanced topics associated to fiber-reinforced polymer composites, especially that related to manufacturing process, with particular reference to liquid composite molding process.

Further, the authors sincerely believe that this book become an excellent reference source for professionals and a start point for encouraging different people to study polymer composite materials, and also fluid flow in porous media.

1.2 What Is a Composite Material?

Composites are materials obtained by a combination (macroscopic level) of two or more distinct physical phases, where at least one part characterizes the reinforcement, load or aggregate, depending on its size and shape (discontinuous phase). The reinforcement is dispersed and embedded in the matrix phase (continuous phase) [1, 2].

There are three main points related to these materials:

(a) They are formed by components with different mechanical and physical properties.
(b) They can be produced by mixing their components. Thus, the dispersed phase can be controlled to obtain suitable properties.
(c) The resulting properties of the components mixing must be superior to those of the individual materials.

Because of the good characteristics presented by the composite materials, like: better performance, safety, economy, durability and versatility, they are being used in large scale when compared to traditional materials [3].

According to chemical and physical nature of the matrix, it is possible to classify composite materials in polymer, ceramic or metallic matrix composites. Based on the chemical structure, the polymers are classified according to the arrangement of the functional group present in the macromolecule, for example polyamides and polyesters. Further, according to preparation method, these materials are classified as addition polymers, which are those formed from a single monomer through an addition reaction without presenting by-products, and condensation polymers, which are formed by different monomers through a condensation reaction with by-products [4].

Besides, based on the technological characteristics, polymers can be classified in thermoplastics, which have linear or branched chains and when heated become malleable being easy to process, and thermosets that have three-dimensional or cross-link structure. The cross-link polymeric structure after cooling cannot back to the original form, becoming insoluble and infusible [5–7].

Now, consider a specified composite material under a stress field. Then, we state that, depending on external variables and their structure with low molecular mass, these materials can exhibit mechanical behavior in two ways after removal of external stresses: elastic solid (Hookean), when it returns to its initial form, and viscous liquid (Newtonian), when the deformation is irreversible. These viscoelastic mechanical behaviors depend on the temperature and time of permanence of the stress acting in them. The viscoelastic deformation disappears completely after a gradual reduction of the stress in a certain time interval. If the applied load exceeds the elastic limit, permanent deformation is obtained [8].

Based on the discontinuous phase, in particular, the reinforcement, composites can be classified in fibrous or non-fibrous, organic or inorganic, natural or synthetic. Further, considering the different types of the existing matrix, the composites can be classified filament, fibrous or particulate, according to the type of arrangement of the discontinuous phase. Then, they can be in the form of short and random fibers; short and aligned fibers; long and aligned fibers; long and random fibers; long and short fibers (randomly oriented), and/or particles.

Based on the classification of the discontinuous phase reported before, we can write [6, 7].

Due to their geometry, these materials can be:

(a) Particulate, tending to the spheroidal shape.
(b) Laminates, in the form of plates with very thin thickness or blanket.
(c) Fibrous, which have a good length/diameter relationship.

Due to their nature, these materials can be:

(a) Extremely hard, to provide increased hardness and abrasion resistance.
(b) Resistant to rupture, to promote high tensile, flexural and shear strength.
(c) Rigid, to promote increase in modulus of elasticity.
(d) Extremely flexible, which increases impact resistance.
(e) Extremely temperature resistant, increasing thermal stability.

In general, fibers are classified based in their origins and physical characteristics. Chemically they can be classified as natural (organic and inorganic), mineral or artificial. When the reinforcement is fibrous, the fibers can be arranged in parallel bundles, so as to form and guide the reinforcement in a preferred direction, multilayer, or in the form of isolated layers or blades. The composites obtained with multidirectional reinforcement have as starting point the textile preforms and constitute a technological leap in order for obtaining massive structures of great volume and with specific properties adjusted to the intended application [3]. The alignment of the reinforcement can facilitate the fabrication of a specified composite, especially in the case of fibers, since they can be provided in the form of fabric, non-woven, wicks, yarns, etc. [9].

One important fact to be considered in the use of a composite material is related to their properties. The physical, electrical and mechanical properties of a composite depend on many factors such as: (a) properties of the constituent materials (fiber and matrix); (b) orientation of the reinforcement; (c) volumetric fraction of the phases (ratio between the volume occupied by a phase and the total volume of the composite); (d) geometry of reinforcement (shape and dimensions), and (e) adhesion between these components.

The interface between the reinforcement and the matrix is one of the most important factors to be considered in composite materials, with one or more continuous phases. The interface corresponds to a hypothetical plan with zero thickness, linking the reinforcement and the matrix by a specific mechanism of adhesion. This region can be considered as one region which the stresses and deformations transmitted by the loading are passed on to the fibers through the matrix [10–12]. The, the final properties of the composites depend strongly on how the individual components interact with each other. In practice, if the adhesion is not perfect, the process of rupture is generated at the interface of the matrix and the reinforcement due to the weakness of the atomic or intermolecular bonds in this region.

For obtaining an optimized interface some physical and chemical factors must be considered such as, the chemical and the physical bonds, the electrostatic and the mechanical interactions, wetting properties, and the phenomenon of contraction that the polymer matrix undergoes during curing process.

More details about this important topic can be found in the literature [6, 7, 13–17].

Anisotropy is the major problem that affects the composite properties. For example, composites reinforced with unidirectional fibers present values of mechanical strength, stiffness, thermal expansion and conductivity different in the directions parallel and perpendicular to the packing of the fibers. To minimize this problem and to improve the composite performance, it is recommended to use laminated composites, where sheets are distributed in a certain sequence [18].

Further, polymer generally has high resistance to chemical and environmental agents and the fibers have high mechanical strength and modulus of elasticity, so, the fiber/polymer joint result in a material with good mechanical properties, high resistance to the chemical agents, increase in stiffness and mechanical resistance, protection against degradation and attack of external agents, increase of thermal conductivity, among others specific characteristics.

Table 1.1 Typical chemical composition of glass fibers (percentage)

Fibrous material	SiO_2	Al_2O_3	B_2O_3	MgO	CaO	Na_2O
Glass E	55.2	14.8	7.3	3.3	18.7	–
Glass C	65	4	5	3	14	8.5
Glass S	65	25	–	10	–	–

In general, fibers have good effectiveness because they present few defects with respect to their shape. They have small diameter and great length, providing a high surface area/volume relationship and, consequently, an increase in the fiber/matrix interfacial area available for the transfer of stress per unit volume of the fiber. There are different types of synthetic fibers, however, glass, carbon and polyaramide fibers can be highlighted. The glass fibers are the most used in composites [6, 19, 20], mainly due to their lower cost, high tensile strength and high chemical inertia, but they present some disadvantages such as: low modulus of elasticity, high density and abrasiveness, and low fatigue strength. Typical compositions for glass fibers are shown in Table 1.1 [21].

On the other hand, natural fibers present various advantages when compared to synthetic fibers, such as: biodegradability, low cost, low density, good tenacity, good thermal properties and reduced use of instruments for their treatment or processing.

In the selection of adequate components for manufacture of fiber-reinforced polymer composites, the choices include not only resin and fiber but also the type of treatment applied at the surface of the reinforcement.

There are a great variety of the vegetable fibers. In this sense, pineapple, banana, coconut, carotid, jute, ramie and sisal fibers have been used as reinforcement in polymeric matrix. One of the main reasons for the excellent properties of the composites reinforced with vegetable fiber results from the fact that these fibers present higher mechanical properties that are much higher than those presented for similar materials.

Mineral particles have been used as filler (loads) for polymer matrix for many years. There are different types of loads in nature, such as limestone (calcite, dolomite), phyllite, mica (muscovite, phlogopite, biotite), silica (quartz, zeolite), kaolin, talc, pyrophyllite (agalmatolite), plaster, barite, wolastonite, smectite (bentonite, montmorillonite, hectorite) and illite [22].

The adequate choice of specific filler depends on the composite performance and surface appearance. These materials induce an increase in stiffness, hardness, strength, electrical properties and chemical resistance, reduction of shrinkage of the molded product, change in density, alteration of color and reduction of flammability, and in some cases, provide an increase in thermal distortion temperature by improving the adhesion between reinforcement and matrix. Further, they must have low moisture content (maximum of 0.1%), low resin demand or oil absorption, low abrasiveness and low cost [23]. Figure 1.1 illustrates a typical $CaCO_3$ mineral load.

Fig. 1.1 Calcium carbonate sample

1.3 Polymer Composite Manufacturing Processes

Composite materials have been experiencing a growing demand in recent years in several sectors of the industry, that is, the search for new processes, which are cheaper and cause a lower environmental impact, is being developed.

One of the main points for the success of material processing is the cost and the reliability of the processing technique employed. For example, metallic materials processing requires machining thinning to suit geometry, but, for polymer composite materials, the processing is done in such a way that at the end of the process a solid with a geometry already in use conditions, thus, the need of a post-process machining is minimal, minimizing costs. There are different manufacturing techniques of composite materials. For thermoset resin-based composite, the following manufacturing processes can be cited [6, 7, 21].

1.3.1 Hand Lay-up Molding

This is the oldest, simplest, and slowest manual production process for manufacturing of polymer composite materials. This can be adapted for the production of large structures, although it is more interesting for small productions [24].

Hand lay-up continues to be one of the most important technique for small batch production. This process is accomplished by arranging the reinforcement in predetermined orientations on the mold. It starts by the mold surface treatment, with the application of a release agent, to facilitate demolding upon resin curing process. Each reinforcing layer is impregnated with the resin preformulated with the catalyst. The thickness of the molded composite is obtained by the number of overlapping layers. The curing is done at room temperature or in an oven, depending on the type of resin used or the need for curing. The curing process, is a reaction between the catalyst and the resin, the reaction hardens the fiber reinforced resin compound without heat production.

The hand lay-up process is performed manually with the help of rollers or brushes by pressing the layers to eliminate bubbles and air and excess resin. The layers of fabric, blanket, or fiber are stacked and impregnated with resin one by one in an open mold. After stacking all layers, the application of an upper molding plate allows a better finish on the outer surface of the composite. Despite the steady progress to replace the manual layout with the automated process, it still persists as the method by which at least half of the entire advanced composite aerospace structure is made [25–27].

The process continues to be used because it is extremely flexible and capable of creating a wide variety of shapes [28]. Although it presents some disadvantages such as the quality of the final product depends on the ability of the worker, the low productivity and the release of volatiles.

1.3.2 Spray-up Molding

Spray up is a process that can be considered an automatic hand lay-up process, being more appropriated for production from medium to large parts and from simple to complex shapes. The process consists in use a spray gun to project the reinforcement (short fiber) and the resin, wrapping the entire surface of the mold where the manufacture of the composite is desired. In this process, the fibers are randomly distributed in the composite formatted. The final thickness of the composite is obtained after several successive applications of the sprayed gun into the mold. After the projection, the composite must be carefully rolled to eliminate trapped air and ensure a good consolidation [24, 29].

The spray up process can be combined with the hand lay up to produce a specific composite structure, since this technique cannot provide the same percentage of fiber as the first, influencing the mechanical properties of the part [24].

1.3.3 Vacuum Bag Molding

The vacuum bag composite molding process is an improved method of the hand lay-up and spray-up molding processes. In this process, the quality of the component improves, due to a greater efficiency in the removal of air bubbles that possibly affect the mechanical properties of the part.

The vacuum bag molding process can be used to remove excess resin from hand lay-up molding process. This process is more conventionally used in the manufacture of high-quality laminates for the aerospace industry using pre-preg materials [30].

1.3.4 Autoclave Molding

In the autoclave molding, the pre-preg is transformed into composites by stacking and compacting the blades forming it. This compaction can be realized in both presses and autoclaves, by using gaseous medium as the physical means of compaction. The composite molding in a press presents limitations as to the size of the mold that the press comprises. In autoclaves operation pressure is carried out by inert gas, which raises the cost of production.

1.3.5 Compression Molding

In this process a mold constituted of two parts is used in the composite manufacturing. The composites molded by compression can be obtained by molding the appropriately formulated reinforcing and matrix mixture. The process starts by the disposition of the reinforcement with appropriate orientation in the mold, treated previously with release chemical agent (demolding agent). The closing of the mold is processed by lowering the top of the press resulting in the consolidation of the material by applied pressure.

The material into the mold is subjected to heat (whenever necessary) and pressure simultaneously, using a hydraulic press with heating. Sufficient amount of material must be placed so that the cavity becomes full with this material. As the mold closes, under pressure, the material is compressed between the two halves, and the excess material flows out the mold as a thin film [31]. After the mold is cooled and the product is removed.

1.3.6 Filament Winding

In the filament winding process, the filament cables are suitably impregnated by polymer resin and subjected to winding on a rotating mandrel. The fibers are unwound under controlled conditions, impregnated into resin and rolled in several layers around a rotating mandrel [27].

A large number of rovings are pulled in a bath containing resin, catalyst and additives. The tension in the fiber is controlled using the fiber guides located between each creel and the resin bath. The excess of resin is removed from the rovings and after winding, the polymerization process (curing process) begins in the oven or through infrared lights [32]. The cure may proceed at room temperature, but in many cases a post cure stage at elevated temperature is necessary depending on the formulation used.

Many variations of this process have been developed, as on-line wetting and prepeg wrappins [28]. The main advantages of this process are the simplicity, automation and the capability to produce parts with good mechanical characteristics. The winding process requires high initial investment into equipments and tooling, but the materials and labor used in the manufacture have low impact in the final cost of the component to be obtained. It is very used in the manufacture of cylindrical parts that will be subjected to internal pressures.

1.3.7 Pultrusion

The pultrusion process starts by the reinforcement impregnation, in the form of continuous fibers, with the previously formulated resin. The most important characteristic is the suitability of the resin gel time, which must be controlled to avoid the premature cure of the resin when it enters the mold. After the impregnation, the fiber/polymer matrix assembly is introduced into the mold which imparts a defined geometry. Ovens are usually used to ensure a perfect cure and increase the velocity of the process. The reinforcement is unidirectional, long and oriented in the direction of the resin flow.

This process has been used with a large variety of resin and fibers types, to produce continuous prismatic sections such as profiles and beams. Pultruded sections offer advantages such as high strength and rigidity as well as a high degree of automation. The main disadvantage is related to the limited geometry [30].

1.3.8 Liquid Molding

The Liquid Composite Molding (LCM) family comprises a set of composite materials manufacturing techniques where a thermosetting liquid resin is injected into an enclosed mold containing a dried fibrous preform impregnating the reinforcement. Among the liquid molding processes, we can cite the RTM (Resin Transfer Molding), infusion, RTM Light, SRIM (Structural Reaction Injection Molding), VARTM (Vacuum-Assisted RTM) and SCRIMP (Seemann's Composites Resin Infusion Molding Process). Following will be presented a discussion about these techniques.

Infusion In this process, the reinforcement is placed into the rigid mold with a removable fabric layer, used to reduce the filling time, along with a layer of fiber to facilitate the flow. Intakes and exits are placed and the vacuum bag is sealed in the mold using a sealant tape. The air is drawn from the cavity to compact the preform between the vacuum bag and the mold, and to subject the resin to a pressure differential. This way, the resin flows through the compacted preform. Infusion is a resin infiltration process within the closed mold that exhibits much similarity with RTM. The most obvious difference, however, is that two rigid mold parts are used in RTM, while in infusion there is rigid part sealed with a vacuum bag.

Resin Transfer Molding (RTM) The RTM process is constituted by three basic steps: pre-processing, processing or injection and post-processing. In the pre-processing, the preform or reinforcement in the format of the part is initially prepared and settled up into the mold. Then, the mold is closed and the preform is compressed. In the mold preparation, before the fibers positioning inside the cavity, it is coated with a release agent so that, after curing process of the resin, the finished composite can be removed without any damage to mold structure.

RTM Light The RTM Light process is gaining global space. The difference between RTM and RTM Light processes are related to the resin injection. In the RTM process it is done through discrete points, while in RTM Light the resin is injected with the aid of an edge or channel (without fibrous reinforcement), which circumvents the entire perimeter of the mold. In relation to the injection pressure, in the RTM Light the vacuum is used, while in the conventional RTM is used a positive pressure to force the resin flow. Thus, the RTM Light process is similar to vacuum assisted molding where the resin is injected at low pressure (~0.5 bar). The use of low pressure allows the use of lighter molds, with little structuring, being quickly and easily constructed. Generally, the mold suffers little deformation and has high durability due to the characteristics of the process. The counter-mold is also semi-rigid and its low weight makes it easy to handle. Further, translucent counter-molds can be used to observe the progress of the resin in the mold. Both the mold and the counter-mold may be made of reinforced plastic with the use of polyester or epoxy resin.

The RTM and RTM Light processes emit little styrene in the environment, and obey a system of standardization and regularity of production, practically isolating the worker from contact with the resin and allowing extreme flexibility of use, in small and large structural or in sandwich material [33].

According to Porto [33], the raw material used in the process depends on the requirements of the part, and the reinforcements may be chopped strands, preforms and fabrics. The resin must have two basic characteristics: low viscosity at room temperature and high reactivity. The waste of raw material when compared to other processes is very low, around 2–7%, due to the burr of the parts. For comparison, in the manual processes, the waste can reach 40% [23]. As in the traditional RTM process, the parts have good finish on both sides and allow large-scale production.

Structural Reaction Injection Molding (SRIM) The SRIM is a process that uses a two-component resin system, which were combined and mixed together. After, this mixture is injected in the mold cavity, where the fibrous reinforcement composed of long fibers in the form of continuous fiber blanket or fabrics is placed. This way, the resin reacts quickly and there is the cure, forming part of the composite [34]. The RTM process uses a static mixer and the resin flows into the mold occurs at a slower rate than the SRIM process (minutes vs. seconds). Further, in the RTM process the filling pressure of the mold is often less than 100 psi, and the required force to fix the mold is also low. In the SRIM process, the mold filling pressure is generally much higher than the pressure used in the RTM process.

Vacuum-Assisted Resin Transfer Molding (VARTM) Vacuum-Assisted Resin Transfer Molding (VARTM) is a process of manufacturing polymer matrix composites in open molds. It is widely used in a variety of commercial applications (e.g. boats, boat hulls, etc.). Furthermore, the VARTM process is applied in different industries, such as: automotive, aerospace and military. The process relies on the use of a single rigid mold (iron plate) which is placed upwardly with reinforcing fiber preforms and enclosed in an air impermeable vacuum bag. The resin is infiltrated in the preform using a vacuum pressure. Finally, the rigid mold is heated to a specified temperature (depending on the resin) and held at that temperature for a period of time sufficient to ensure complete cure of the prepreg with resin [35].

The VARTM process offers several advantages over other manufacturing processes of polymer matrix composites, such as: (a) low processing cost, (b) low level of emissions of volatile organic chemicals, (c) processing flexibility, (d) low content of voids, and (e) a potential for relatively large volume manufacturing (surface area ~150–200 m^2) and the thickness of the composite part (0.1–0.15 m), containing a high content of reinforcing fiber (75–80% in weight).

Seemann's Composites Resin Infusion Molding Process (SCRIMP) In the Seemann's Composites Resin Infusion Molding Process (SCRIMP), a mold is used on only one side of the piece and the vacuum pressure for infusion of the resin is used. The fibrous reinforcement is placed in the mold soon after the release agent is applied. Following, the reinforcement is placed on a thin layer of material that does not adhere to the resin to facilitate release of the part. Finally, a dispensing medium is placed. That dispensing medium is, in fact, a high permeability material, that causes an increase in the speed of the resin at the surface and, consequently, allows resin flow in the transverse direction and, thereafter, the vacuum bag is placed. An advantage of this process is the use of only one side of the mold, which reduces the cost. Disadvantages include the difficulty of automating the process and the fact that the side of the part molded by the vacuum bag does not have a good surface finish [36].

References

1. Mano EB, Mendes LC (1999) Introduction to polymers, 2nd edn. Edgard Blucher Ltda, São Paulo, Brazil (In Portuguese)
2. Carvalho RF (2005) Composites of sisal fibers for use in reinforcement of wood structures. Doctoral Thesis, Sciences and Materials Engineering, University of São Paulo, São Paulo, Brazil. (In Portuguese)
3. Idicula H, Boudenne A, Umadevi L, Ibos L, Candau Y, Thomas S (2006) Thermophysical properties of natural fibre reinforced polyester composites. Comp. Sci. Technol. 66(15):2719–2725
4. Callister WD Jr, Rethwisch DG (2008) Fundamentals of materials science and engineering: an integrated approach, 3rd edn. Wiley, Hoboken, USA
5. Alexandre MEO (2005) Composites of polyester matrix reinforced with pineapple fiber: mechanical properties and water absorption. Doctoral Thesis, Federal University of Rio Grande do Norte, Natal, Brazil. (In Portuguese)

6. Santos MJN, Delgado JMPQ, Lima AGB (2017) Synthetic fiber-reinforced polymer composite manufactured by resin transfer molding technique: foundations and engineering applications. Diffus Foundations 14:21–42
7. Melo RQC; Lima AGB (2017) Vegetable fiber-reinforced polymer composites: fundamentals, mechanical properties and applications. Diffus Foundations 14:1–20
8. Engel L, Klingele H, Ehrenstein GW, Schaper H (1981) Atlas of polymer damage. Prentice Hall, London, England
9. Mitchell BS (2004) An introduction to materials engineering and science. New Jersey, USA
10. Cavalcanti WS (2006) Composites polyester/woven plant-glass fabrics: mechanical characterization and simulation of water sorption. Doctoral Thesis, Process Engineering, Federal University of Campina Grande, Campina Grande, Brazil. (In Portuguese)
11. Razera IAT (2006) Lignocellulosic fibers as a reinforcing agent for phenolic and lignofenolic matrix composites. Doctoral Thesis, University of São Paulo, Institute of Chemistry of São Carlos, São Carlos, Brazil. (In Portuguese)
12. Chawla KK (2012) Composite materials: science and engineering. Springer-Verlag, New York, USA
13. Callister Jr WD (2003) Materials science and engineering: an introduction, 6th edn. Wiley, New York, USA
14. Kim J-K, Mai Y-W (1998) Engineered interfaces in fiber reinforced composites. Elsevier Science Ltd., Oxford, England
15. Kaw AK (2006) Mechanics composite materials, 2nd edn. CRC Press, Boca Raton, USA
16. Hull D, Clyne TW (1996) An introduction to composite materials. Cambridge Solid State Science Series, 2nd edn. England, Cambridge
17. Matthews FL, Rawlings RD (1994) Composite materials: engineering and science. Chapman & Hall, Oxford, United Kingdom
18. Araújo JR (2009) High density polyethylene composites reinforced with carauá fibers obtained from extrusion and injection. Master Dissertation. State University of Campinas, Institute of Chemistry, Campinas, Brazil. (In Portuguese)
19. Shackelford JF (2005) Materials science for engineers, six edn. Peason Prentice Hall, New Jersey, USA
20. Callister WD Jr (2008) Materials science and engineering: An introduction, 7th edn. Wiley, New York, USA
21. Levy Neto F, Pardini LC (2006) Structural composites—science and technology, 1st edn. Edgard Blucher, São Paulo, Brazil
22. Lewin A, Mey-Marom A, Frank R (2005) Surface free energies of polymeric materials, additives and minerals. Polym Adv Technol 16(6):429–445
23. ABMACO, Brazilian Association of Composite Materials. Available at: http://www.abmaco.org.br/, 15/04/2013
24. Bunsell AR, Renard J (2005) Fundamentals of fibre reinforced composite materials. Institute of Physics (IOP), UK
25. Castro BFM (2013) Study and mechanical characterization of composites reinforced with natural fibers. Master Dissertation. Mechanical Engineering, University of Porto, Porto, Portugal. (In Portuguese)
26. Astrom BT (1997) Manufacturing of polymer composites. CRC Press, Boca Raton, USA
27. Moura MFSF, Morais AB, Magalhães AG (2010) Composite materials—materials, manufacturing and mechanical behavior. Publindústria, Porto, Portugal (In Portuguese)
28. Gutowski TG (1997) Advanced composites manufacturing. Wiley, Cambrigde, USA
29. Mazumdar SK (2002) Composites manufacturing: materials, product and process engineering. Taylor & Francis, New York, USA
30. Rudd CD, Turner MR, Long AC, Middleton V (1999) Tow placement studies for liquid composite moulding. Comp Part A 30(9):1105–1121
31. Ratna D (2009) Handbook of thermoset resins. Smithers Publishers, Shawbury, United Kingdom, pp 19–21

32. Mallick PK (2007) Fiber-reinforced composites: materials, manufacturing and design. CRC Press, New York, USA
33. Porto JS (2012) Computational modeling of the resin transfer molding process applied to naval engineering. Master Dissertation. Computational Modeling, Federal University of Rio Grande, Rio Grande, Brazil. (In Portuguese)
34. Lin MY, Kang MK, Hahn HT (2000) A finite element analysis of convection problems in RTM using internal nodes. Comp Part A 31(4):373–383
35. Grujicic M, Chittajallu KM, Walsh S (2003) Optimization of the VARTM process for enhancement of the degree of devolatilization of polymerization by-products and solvents. J Mater Sci 38(18):3729–3739
36. Alves ALS (2006) Processing of composite plaques by the SEEMANN resin infusion molding technique (SCRIMP). Master Dissertation. Federal University of Rio de Janeiro, Rio de Janeiro, Brazil. (In Portuguese)

Chapter 2
The Liquid Composite Molding Process: Theory and Applications

Abstract This chapter focuses on the liquid composite molding technique with special attention to resin transfer molding process (RTM). Herein, the main issues related to this manufacturing technique such as foundations, processing stages, main control variables, problems and advantages associated with the use of the RTM technique, and finally applications of fiber-reinforced polymer composite at different areas are presented and discussed.

2.1 Foundations

The Resin Transfer Molding (RTM) is a process that consists basically of three steps: pre-processing, processing or injection and post-processing. In pre-processing, the preform or reinforcements in the shape of the part is initially prepared and positioned in the mold. Thereafter, the mold is closed and the preform is compressed. In preparing the mold, before positioning the preform into the cavity, it is coated with a release agent so that after curing the resin, the finished composite can be removed without any damage to its structure. The definition of injection pressure is another important step in RTM processing. Usually are used pressures that are between 1 and 10 bar, because larger values can cause: (a) drag of reinforcement; (b) deflection of the mold, and (c) formation of voids in the composite. In cases where the structure has large dimensions, the mold should be reinforced or a press can be used so that there is no deflection of it during the injection process of the resin.

In the second step, of injection or processing, the resin fills the mold and the fibrous reinforcement is impregnated. At that moment the monitoring of the process must be done, and the verification that it is developing properly is crucial. Examples of process parameters to be monitored are the injection pressure, the resin injection enters points and the resin exit points, among others. The mold must have at least one resin injection port and an outlet part to allow, during resin injection, the outlet of the air inside the mold. In larger pieces, it is common, however, the presence of several input and output gates [1, 2]. The resin to be used in the process must have a low viscosity ($\mu \leq 1$ Pa.s) so that a good impregnation of the reinforcement is obtained and the injection time is not very high. Finally, in the last step or post-processing, curing of

J. M. P. Q. Delgado et al., *Transport Phenomena in Liquid Composite Molding Processes*, SpringerBriefs in Applied Sciences and Technology,
https://doi.org/10.1007/978-3-030-12716-9_2

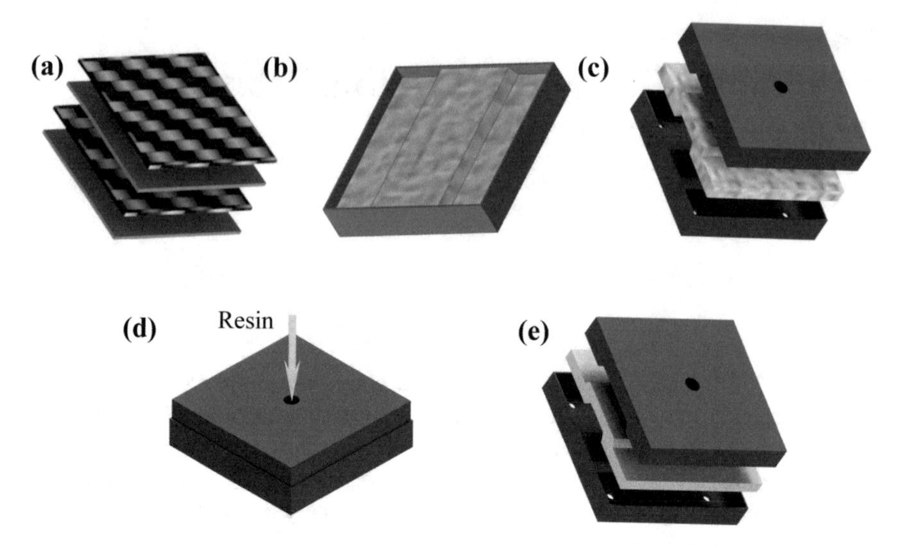

Fig. 2.1 Basic steps of the RTM process; **a** Preform; **b** Preform in the mold; **c** Closing mold; **d** Closed mold, resin injection and cure process, and **e** Demolding and final processing

the resin occurs (in situ, with mold at room temperature or heated) and the demoulding process. Following, the composite can undergo a treatment with chemical products and a polishing process, in order to leave it ready for commercialization, reaching optimized properties [3]. Figure 2.1 schematically illustrates the steps related to the RTM process.

According to Amorim Jr. [3], open mold operations may take a significant amount of time and the closed mold operations are actually those that have the potential to control the rate of the production, cost and performance of the part produced. However, to obtain the maximum of the mechanical properties of the composites manufactured by RTM, each individual fiber must be completely "wet" by the liquid resin and the voids spaces must be as low as possible. The wettability of the fibers depends on the material properties, such as superficial energy of the resin/fiber system and its time of contact. In addition to the wettability problem, macro-voids generated by the coalescence of the flow front may occur; therefore, the dynamics of the fluid inside the mold is very important for the performance of the RTM processing. Studies related to resin rheology, mold geometry, resin/fiber interaction, positioning of resin injection points, air outlet points, etc., are essential for optimization of the process. Then, for obtain a good RTM molding, it is necessary to know and control different processes parameters, such as the viscosity, injection pressure, fiber volumetric fraction, porosity and, the temperature and permeability of the medium.

Porosity can be defined as the fraction of the total volume of the porous media that is occupied by pores connected and disconnected [4, 5]. However, only the connected pores are available to fluid motion relative to the solid skeleton. In RTM process, this parameter is dependent of the quantity, size and shape of the reinforcement.

Permeability is a measure of how easily a single-phase fluid moves in a porous medium under the influence of a dynamic pressure gradient (absolute pressure gradient adjusted for gravitational effects) [6].

However, when two or more immiscible phases (multiphase fluid system) exist into the pore space (which fills out the voids completely), they are separated by interfaces where interfacial forces act producing discontinuities in density and pressure [7]. Then, fluid motion is characterized by different velocities between the phases. Thus, appear a new concept to phase permeability, so called relative permeability of the phase, which in turn, is dependent of the local fluid saturation. Further, we state that the vegetable fiber act as fluid sinks, so, fluid velocity is reduced thus, altering local pressure and fluid saturation. Then, the pressure drop produces an apparent change in the permeability along the preform.

For this physical situation as applied to RTM process, we can cite, air and resin (two-phase system), and air, resin and particulate load (three-phase system), stopped or flowing through a porous media.

Particulate load can alter resin viscosity and density, and porous media permeability, so, fluid velocity is dependent of the different parameters simultaneously, which difficult the analysis of this process parameter. Thus, we can define a new parameter that incorporate all these parameters, so called phase mobility. This parameter is a relationship between fluid phase permeability and viscosity [8].

The injection pressure and the temperature gradient are the main factors that affect the correct filling of the mold and thus the manufacturing productivity. Other important variables to the RTM process development are: amount and location of the injection gates and mold geometry [9–11].

The pressure and the injection velocity of the resin can induce failures such as preform deformation at the time of the injection (fiber wash out), formation of dry spots in unpredictable places and the phenomenon of race tracking, preferred route chosen by the resin during the injection, occurring mainly in regions with smaller amounts of fibers. The definition of injection pressure is another important step in RTM processing. Pressures that are between 1 and 10 bar are commonly used because larger values can provoke fiber drag, deflection of the mold/counter-mold, and formation of voids in the mold.

The micro-voids are another type of defect, generally occurring due to the lower permeability inside the fibers. Unlike the surface and region between the fibers, the internal region of the fibers has its permeability governed by the capillary pressure [12, 13].

A major concern of RTM process is the temperature distribution during mold filling because the temperature of the injected resin differs from that of the mold creating a temperature gradient in the mold cavity which consequently promotes uneven cure in the part. This temperature gradient also occurs in the thickness of the material, a phenomenon that can cause a change in resin viscosity and thus, changes in the resin front that impregnates the preform [14, 15].

Another factor to consider is the low thermal conductivity of the composite, especially when they are processed with larger thickness. Because this parameter generates residual stresses inside the material after processing. This fact can be attributed only to the contraction (in some cases, about 9%) of the resin during the polymerization (curing process). A solution for induced residual stresses is to decrease the processing temperature in such a way that it does not affect the polymerization reaction [1, 15, 16].

In RTM processes involving polymers with exothermic reaction there is always heat exchange between the preform and the heated mold, characterizing a non-isothermal injection process. The influence of the heat generated in the polymerization reaction during the injection can be related to the point of gelation, which relates the injection time to the time required for the start of the reaction. It is observed that the lower the point of gelation index, the less influence the reaction time has on the process [16].

The knowledge of the permeability of the preform plays a fundamental role in the quality and reproducibility of the pieces associated to the quantity and distribution of voids throughout the laminate. The inadequate flow of resin induces voids and dry spots due to the unbalanced flow associated with the permeability variation of the preform. There are several variables that control permeability and are important for processing by RTM [1, 3], such as:

(a) Preform fabric architecture.
(b) Defects in fibers (cracks, etc.).
(c) Type of fiber.
(d) Sizing of fibers.
(e) Compressibility of fibers.
(f) Number of layers of fabric.
(g) Injection parameters (resin flow rate and injection pressure).
(h) Porosity (micro and macro porosities).
(i) Capillarity (interaction of the resin with the fiber).
(j) Volumetric fraction of the fibers in the composite.
(k) Test fluid used in the permeability study.
(l) Mold deflection.
(m) Wettability.

2.2 Advantages and Disadvantages of the RTM Technique

The Resin Transfer Molding (RTM) process has proved to be very effective in the processing of high-performance composite materials, and allows the fabrication of complex structures with mechanical quality and good surface finishing.

Some advantages that can be mentioned of the RTM process in comparison to other manufacturing processes [1, 17], such as sheet molding compound (SMC) and injection molding, are:

(a) Low labor cost.
(b) Simple tooling.

(c) Manufacture of complex, quality structures.
(d) Parts with two sides with smooth finish (loss reduction).
(e) Production scale.
(f) Parts with different sizes and shapes.
(g) Easy process control.
(h) Reduction (low level) of styrene emission.
(i) Possibility of incorporating inserts, shoulders, ribs and other reinforcements in the molded parts.
(j) Mechanical properties comparable to other processes (assuming the same resin system and equal fiber content).
(k) Reinforcements or blends of reinforcements can be arranged to obtain specific mechanical characteristics.
(l) Load systems can be used to reduce the cost, increase stiffness, increase flame retardancy performance and low smoke emission, improve surface finish and reduce the exothermic peak of the resin to reduce cracking.
(m) Tools for the RTM process generally cost less than other closed mold processes, and in most cases, are constructed of glass fiber reinforced polyester.
(n) One or both faces of the part may have gel-coat.
(o) Veil or surface blanket can be used where it requires better corrosion resistance and/or better surface appearance.
(p) Curing cycles are generally faster than the open molding cycles (satisfactory cycle of time).
(q) Constant thickness and dimensional accuracy.
(r) Accurate reinforcement/resin control.
(s) Cleaning in the mold area.
(t) Minimum finish (burrs) on finished parts.

On the other hand, among several disadvantages, the following can be mentioned:

(a) Invisible for negative output parts.
(b) Complex mold movement.
(c) Models and molds of careful construction and subsequent modification that are not feasible.
(d) Rigid physical-chemical controls of materials and process.
(e) Constant maintenance of equipment and means of production.
(f) Accurate operational training.
(g) High cost of mold acquisition.
(h) Limitation of the dimensions of the mold and the part to be obtained.

One of the greatest benefits of RTM processing is the separation between the molding process and architecture type of the part. In this way, the fabrics can be placed in the desired direction and order in the preform and then placed inside the mold for resin injection and curing. These two distinct phases, arrangement of the preform fabric organization and resin injection/cure, allow the designer to develop materials for specific applications, that is, to satisfy local and specific conditions of the load [3, 18].

2.3 Engineering Applications

After decades of restricted use in some industry sectors, such as missiles, rockets and aircraft of complex geometries, structural polymer composites, also called advanced, have expanded their use in different sectors of modern industry. This is due to the high degree of automation that the recent processing techniques allow. The use of structural polymer composites has presented a growth of 5% per year in average, promoting a considerable energy saving compared to the processing of metallic component (it does not require exhaust systems). These factors are essential to meet project specifications and to reduce operating costs [12].

Nowadays, the use of high performance and low weight structures has been made in the automotive industries (frames, rear and front chassis, hoods and mudguards of vehicles, agricultural machinery, buses and trucks, etc.), sports (frame and handle of bicycles, golf clubs, jet-skis, skateboards, etc.), civil construction (columns and posts), naval (boat hulls, masts, decks, etc.), aerospace and military (fuel tanks, launch tubes, military equipment boxes, etc.), biomedical, among others [18–22].

In the maritime market, generally the resins applied for the production of a composite material are usually the thermosetting polyester, vinyl ester and epoxy. The choice of resin depends on the structural requirements of the composite, ease of handling, curing characteristics, the type of environment where it will be used, the operating temperature and the life time for which the structure is designed. In boats, the propulsion system is a component of great importance. Boat propellers manufactured with fiberglass-reinforced polymers have high corrosion resistance when compared to traditional metal propellers, as well as resisting more abrasion, presents high propulsive efficiency due to low weight and greater fuel economy in operation [1, 23].

References

1. Santos MJN, Delgado JMPQ, Lima AGB (2017) Synthetic fiber-reinforced polymer composite manufactured by resin transfer molding technique: foundations and engineering applications. Diffus Foundations 14:21–42
2. Alves ALS (2006) Processing of composite plaques by the SEEMANN resin infusion molding technique (SCRIMP). Master Dissertation. Federal University of Rio de Janeiro, Rio de Janeiro, Brazil. (In Portuguese)
3. Amorim Jr. WF (2007) Processing of thick composite board through resin transfer molding, Doctoral Thesis, Department of Metallurgical and Materials Engineering, Federal University of Rio de Janeiro, Rio de Janeiro, Brazil. (In Portuguese)
4. Nield D, Bejan A (2006) Convection in porous media, 3rd edn. Springer, New York, USA
5. Bear J (1972) Dynamics of fluid in porous media. Dover Publications Inc., New York, USA
6. McKibbin R (2005) Modeling heat and mass transport processes in geothermal systems. In: Vafai K (ed) Handbook of porous media, 2nd edn. Taylor & Francis, Boca Raton, USA, pp 545–571
7. Harris SD, Ingham DB (2005) Parameter identification within a porous medium using genetic algorithims. In: Vafai K (ed) Handbook of porous media, 2nd edn. Taylor & Francis, Boca Raton, USA, pp 687–742

8. Santos MJN, Delgado JMPQ, Lima AGB, Oliveira IR (2018) Liquid injection molding process in the manufacturing of fibrous composite materials: Theory, advanced modeling and engineering applications. In: Delgado JMPQ, Lima AGB (eds.) Transport phenomena in multiphase systems, 1st edn. Springer International Publishing, Cham, Switzerland, pp 251–272
9. Oliveira IR, Amico SC, Luz FF, Barcella R, Bezerra VMF, Lima AGB (2013) Effect of $CaCO_3$ content in resin transfer molding process. Deff Diff Forum 334–335:188–192
10. Oliveira IR, Amico SC, Barcella R, Lima AGB (2014) Application of calcium carbonate in resin transfer molding process. Deff Diff Forum 353:39–43
11. Pearce NRL, Guild FJ, Summerscales J (1998) An investigation into the effects of fabric architecture on the processing and properties of fibre reinforced composites produced by resin transfer moulding. Compos. Part A 29(1–2):141
12. Brocks T (2011) Carbon/epoxy structural composition via RTM for aeronautical application: processing and characterization. Master Dissertation. Mechanical Engineering, Paulista State University, Guaratinguetá, Brazil. (In Portuguese)
13. Lee DH, Lee W II, Kang MK (2006) Analysis and minimization of void formation during resin transfer molding process. Compos Sci Technol 66(16):3281–3289
14. Shiino MY (2011) Development and characterization of compounds processed by RTM for aerospace application. Master Dissertation. Mechanical Engineering, Paulista State University, Guaratinguetá, Brazil. (In Portuguese)
15. Lebrun G, Gauvin R, Kendall KN (1996) Experimental investigation of resin temperature and pressure during filling and curring in a flat steel RTM mould. Comp Part A 27(5):347–356
16. Ruiz E, Trochu F (2005) Numerical analysis of cure temperature and internal stresses in thin and thick RTM parts. Compos Part A 36(6):806–826
17. Hamidi YK, Aktas L, Altan MC (2005) Three-dimensional features of void morphology in resin transfer molded composites. Compos Sci Technol 65(7–8):1306–1320
18. Goss TM (2010) View and perspectives of the Brazilian composite industry. UnderGraduate Final Work. Department of Chemical Engineering, Federal University of Rio Grande do Sul, Porto Alegre, Brazil. (In Portuguese)
19. Rezende MC, Botelho ECO (2000) The use of structural composites on the aerospace industry. Polímeros: Ciência e Tecnologia 10(2):e4-e10. (In Portuguese)
20. Steenkamer DA (1994) The influence of preform design and manufacturing issues on the processing and performance of resin transfer molded composites. Ph.D. Thesis, University of Delaware, Center for Composite Materials, Delaware, USA
21. Hsião K-T, Little R, Restrepo O, Minaie B (2006) A study of direct cure kinetics characterization during liquid composite molding. Compos Part A Appl Sci Manuf 37(6):925–933
22. Oliveira IR (2014) Infiltration of loaded fluids in porous media via RTM process: Theoretical and experimental analyses. Doctoral Thesis, Process Engineering, Federal University of Campina Grande, Campina Grande, Brazil
23. Fontoura CM, Rocha ICL, Ferreira CA, Amico SC (2009) Characterization of composite materials made by resin transfer molding for application in propulsor systems. In: Proceedings of the 10th Brazilian Congress of Polymers 10th CBPol, Foz do Iguaçu, Brazil, 10 p. (In Portuguese)

Chapter 3
Advanced Experiments in RTM Processes

Abstract This chapter provides experimental information concerned with fluid flow in unsaturated porous media, with particular reference to resin transfer molding. Herein, two important topics will be presented: the rectilinear and radial infiltration processes. Results of the flow front position and injection pressure, with or without fillers, in different instants of infiltration process are shown and analyzed. The study is essential to evaluate the mechanical performance of fiber-reinforced polymer composites and to validate new and advanced mathematical modeling related to fluid flow in porous media.

3.1 The Liquid Molding System

Several experiments were conducted using synthetic fiber mat molded in an RTM system. The experiments were performed in a stainless-steel mold. The mold has the top made of glass to enable viewing of the advance of fluid flow. The system has a camera positioned above the mold for monitoring the impregnation process with the aid of a timer. Figure 3.1 illustrates the diverse components that composes the used RTM system. In Fig. 3.2, the main steps for one RTM experiment are shown.

3.2 Rectilinear Infiltration

Oliveira [1] and Oliveira et al. [2–4] conducted rectilinear injection experiments of orthophthalic polyester resin in a glass fiber mat (450 g/m^2) inserted in a stainless steel mold using a RTM system (room temperature and different maximum injection pressure and $CaCO_3$ content). The mold with cavity dimensions of 320 $\times$ 150 $\times$ 3.6 mm^3, has one inlet and two outlet points (vents), and the top is made of glass to enable viewing of the advance of fluid flow. The resin's viscosity was measured in a Brookfield viscometer HBDV-II + C/P with the S51 spindle, and density of the resin mixed with Calcium Carbonate was obtained by using the mixture rules.

J. M. P. Q. Delgado et al., *Transport Phenomena in Liquid Composite Molding Processes*, SpringerBriefs in Applied Sciences and Technology,
https://doi.org/10.1007/978-3-030-12716-9_3

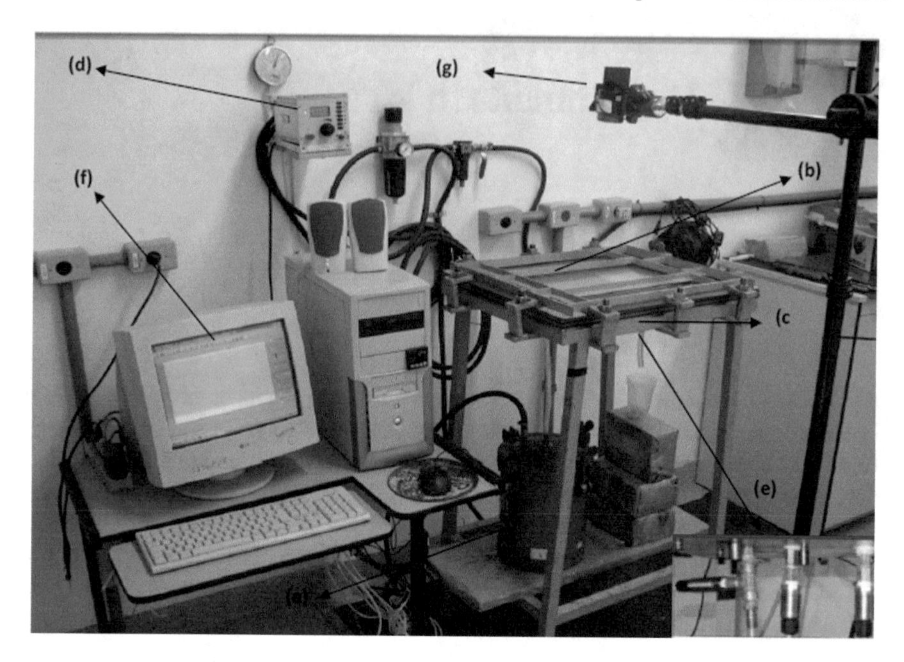

Fig. 3.1 Photo of the RTM experimental apparatus available at LACOMP/Federal University of Rio Grande do Sul (Brazil): **a** pressure vessel, **b** strengthened glass top mold, **c** steel lower mold, **d** pressure controller, **e** pressure transducers, **f** data acquisition system and **g** camera

The RTM equipment is shown in Figs. 3.1 and 3.2. Figure 3.3 shows the stainless-steel mold used in the experiment and Fig. 3.4 shows the preform inserted in the mold.

In rectilinear infiltration, the fluid (resin) is introduced by an input port at the end of the mold at a volumetric flow rate limited by the mold cavity, and flows in the direction of the vent points at the other end.

In the situation where the injection pressure (P_{inj}) is constant, the time (t_{ff}) required to the flow front of the fluid to reach a specific position (x_{ff}), within the mold, can be obtained using the following equation [3–7]:

$$k = \frac{\varepsilon\mu}{2P_{inj}t_{ff}}X_{ff}^2 \tag{3.1}$$

where ε and V_f ($\varepsilon = 1 - V_f$) represent, respectively, the media porosity and volume fraction of the fiber. Since that, injection pressure, time, viscosity, porosity and flow front position are obtained experimentally, therefore the Eq. (3.1) can be used to determining the porous media permeability (k).

The fiber volume fraction inside the mold can be calculated using the Eq. (3.2), as follows:

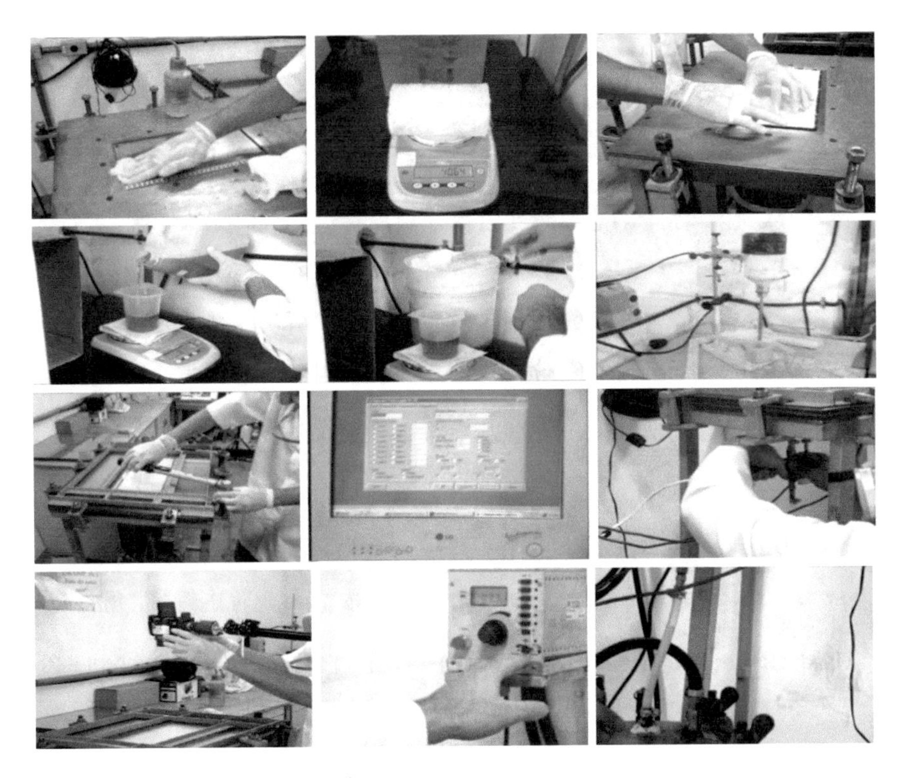

Fig. 3.2 Basic steps for RTM experiment

Fig. 3.3 Stainless steel RTM mold without the preform and the top glass (rectilinear infiltration)

Fig. 3.4 Preform inserted into the mold

$$V_f = \left(\frac{G}{\rho h} n\right) \tag{3.2}$$

where G is the fiber weight, ρ is the fiber's specific mass, h is the mold thickness and n is the fiber mat's stacked quantity.

For determining of the mixture density (resin + $CaCO_3$) was used the following equations:

$$\rho_m = \rho_R V_R + \rho_{CaCO_3} V_{CaCO_3} \tag{3.3}$$

where ρ_m is the mixture density, and ρ_R is the resin density and ρ_{CaCO_3} is the Calcium Carbonate density; V_R represents the volumetric fraction of the resin and V_{CaCO_3} is the volumetric fraction of the $CaCO_3$ [8].

The parameter V_r and V_{CaCO_3} can be determined as follows:

$$V_R = \frac{v_R}{v_{Total}} \tag{3.4}$$

$$V_{CaCO_3} = \frac{v_{CaCO_3}}{v_{Total}} \tag{3.5}$$

where v_r and v_{CaCO_3} represent the volumes of the resin and Calcium Carbonate, respectively, calculated as follows:

$$v_R = \frac{m_R}{\rho_R} \tag{3.6}$$

$$v_{CaCO_3} = \frac{m_{CaCO_3}}{\rho_{CaCO_3}} \tag{3.7}$$

$$v_{Total} = v_R + v_{CaCO_3} \tag{3.8}$$

Table 3.1 Viscosity and permeability data for each experiment

Case	Viscosity (cP)	Maximum injection pressure (bar)	Permeability ($\times 10^{-3}$ m^2)
0% $CaCO_3$	330	0.203	6.6
40% $CaCO_3$	2113	0.213	11.9

In the Eqs. (3.6)–(3.8), the parameters m_r, m_{CaCO_3} and v_{total} are, respectively, mass of resin, mass of $CaCO_3$, and total volume of the mixture resin plus $CaCO_3$.

The following infiltration process conditions were used in the RTM experiments:

(a) Non-filled resin (0% $CaCO_3$), resin density $\rho = 1190$ kg/m^3, fiber volume fraction $V_f = 19\%$, porosity $\varepsilon = 0.81$;
(b) Filled resin with 40% $CaCO_3$, resin density $\rho = 1500$ kg/m^3, fiber volume fraction $V_f = 19\%$, porosity $\varepsilon = 0.81$.

Table 3.1 shows results for viscosity and permeability as a function of calcium carbonate content and injection pressure. It is observed that the addition of $CaCO_3$ modifies the permeability and increases the viscosity. It is realized that the higher the concentration of $CaCO_3$ in the resin, the greater the permeability enhancement due to the inherent characteristics of the experimental method used for estimation of this parameter (rectilinear flow).

We state that the resin flow (with 40% of $CaCO_3$) was interrupted in 1261 s of injection process, due to presence of $CaCO_3$, before the complete filling of the mold (fluid flow lockout).

Figures 3.5 and 3.6 compare digital images of the flow front advancing into the mold for the cases pure resin (0% $CaCO_3$) and resin filled with 40% of $CaCO_3$, respectively, at different elapsed times. Figures 3.7 and 3.8 illustrate the injection pressure behavior and flow front position versus time, respectively, captured during the experiments, for all the studied cases.

In these figures, it is observed that the filling time velocity decreases with the increasing of the $CaCO_3$ content in the resin, consequently, it needs more time to fill the mold. The injection pressure has also its effect on the total filling time, the higher the injection pressure and lower the $CaCO_3$ content less the total filling time.

3.3 Radial Infiltration

Luz [9] and Luz et al. [10, 11] conducted several experiments of resin injection in a porous pre-form (300 × 300 × 20 mm^3). The fluid used in the experiments was a commercial vegetable oil (soybean oil) with density of 914 kg/m^3. The fluid's viscosity at a temperature of 23 °C (same temperature that occurred in the experiments) was 37.1 cP, measured in a Brookfield viscometer HBDV-II + C/P with the S40 spindle. For the fibrous reinforcement, was used woven 0/90 of E-glass fiber from

Fig. 3.5 Experimental front flow in the non-filled resin case (0% $CaCO_3$ and $V_f = 19\%$) at different times

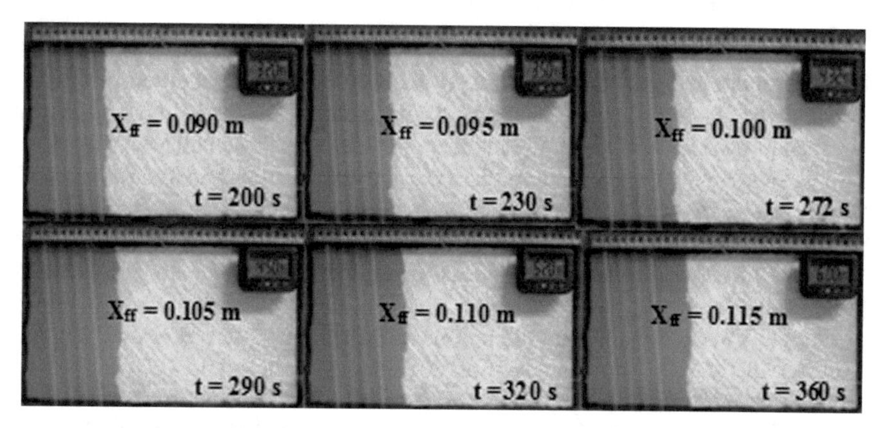

Fig. 3.6 Comparative of the experimental data for the fluid flow front for the filled resin case (40% $CaCO_3$ and $V_f = 19\%$)

Owens Corning (300 g/m^2). In the experiments we use the RTM mold, with radial injection, which equipment schematic is shown in Fig. 3.9. The mold contains one injection port and four outlet ports (vents).

When using radial infiltration in RTM process, the time required (t_{ff}) for the fluid, which passes through an injection port with a radius r_{inj} and injection pressure P_{inj} (gauge), fill a region of radius r_{ff} in the mold is given by Eq. (3.9) [9–12]. This equation is applied only until the flow reaches the wall of the mold.

$$k = \frac{\varepsilon\mu}{2P_{inj}t_{ff}}\left[r_{ff}^2 \ln\left(\frac{r_{ff}}{r_{inj}}\right) - \frac{1}{2}\left(r_{ff}^2 - r_{inj}^2\right)\right] \tag{3.9}$$

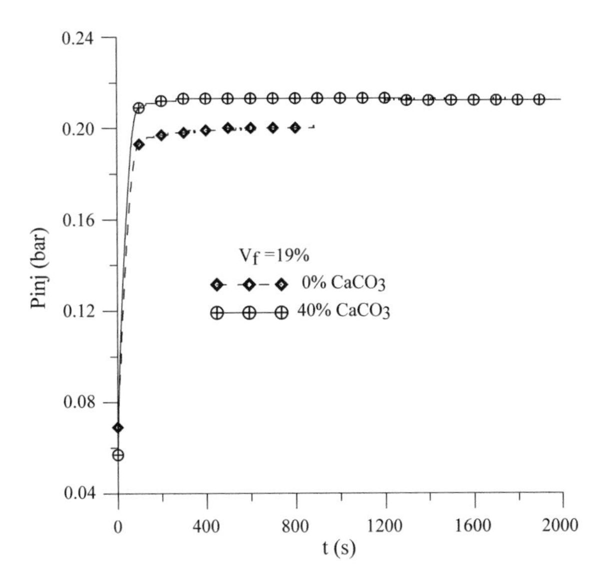

Fig. 3.7 Injection pressure transient (rectilinear flow) behavior

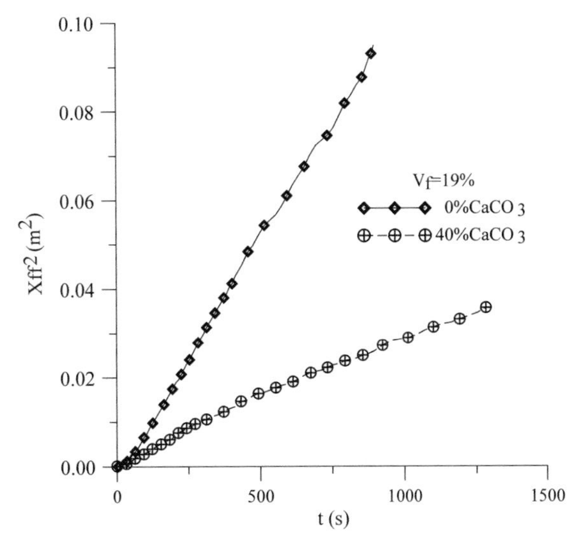

Fig. 3.8 Fluid flow front position as a function of the time (rectilinear infiltration)

Therefore, during the infiltration process, the radius of the flow front was measured at different times and, using Eq. (3.9), it was determined the permeability of the medium. The parameters used in the experiments, and the permeability data found are described in Table 3.2.

Figure 3.10 shows the volume fraction of the injected fluid evolution (Case 2) in three different times (60, 150 and 360 s). Whereas Fig. 3.11 illustrates the distribution of this parameters at the moment where the resin touches the first wall of the mold (t_{border}) for all studied cases (see Table 3.2).

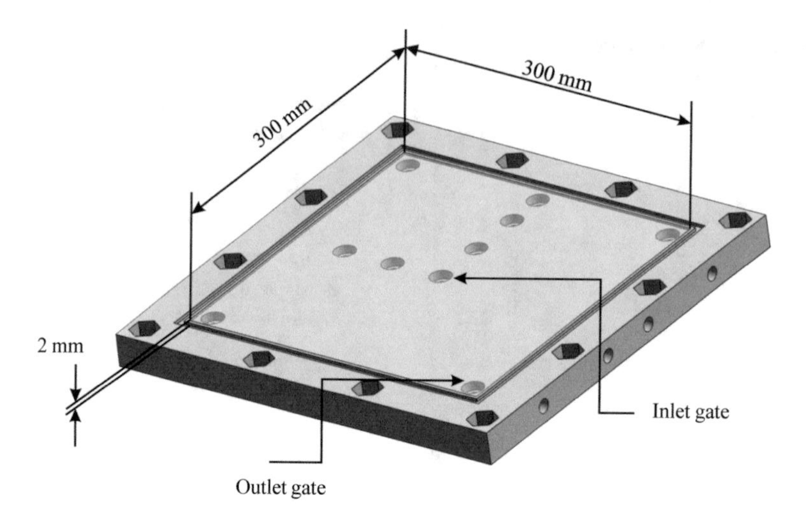

Fig. 3.9 Stainless steel RTM mold without the preform and the top glass (radial infiltration)

Table 3.2 Permeability and mold filling time for several experiments

Case	$V_{f\,real}$ (%)	k ($\times 10^{-11}$ m^2)	t_{border} (s)
1	42.2	2.2	640
2	41.7	2.0	560
3	41.9	3.7	380

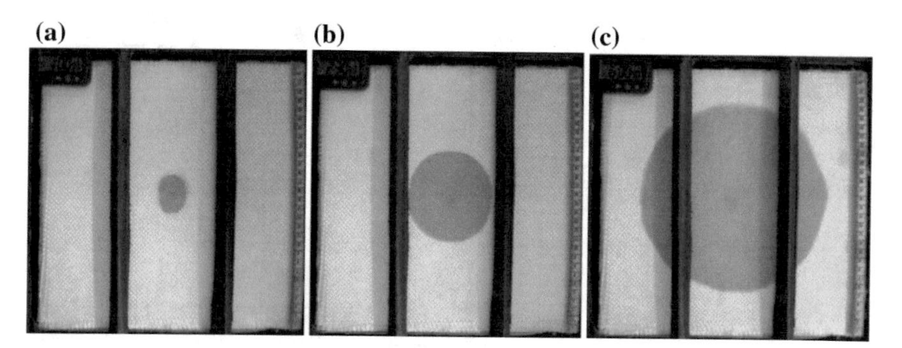

Fig. 3.10 Forward flow evolution of the fluid in the case 2. **a** 60 s, **b** 150 and **c** 360 s

The resin's injection pressure in the mold was monitored during each experiment. From the pressure data collected, was made a non-linear regression using the Mathematica® software resulting in a pressure equation as a function of process time.

The transient pressure behavior is given by Eq. (3.10) obtained by fitting to the experimental data [9–11]:

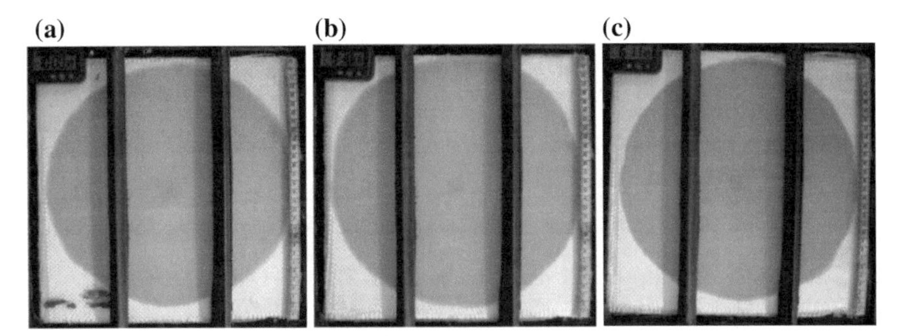

Fig. 3.11 Volume fraction comparison in the t_{border}: **a** Case 1, **b** Case 2 and **c** Case 3

Table 3.3 Parameters of the Eq. (3.10)

Case	P_e (mbar)	t_e (s)	a_1 (mbar/$s^{0.25}$)	a_2 (mbar)	a_3 (s^{-1})	a_4 (−)	a_5 (s^{-1})
1	1679.72	240	30.41	5.29×10^{-1}	−72.28	−178.62	−9.67
2	1888.35	294	45.20	0.18	−3099.14	−6960.16	−352.70
3	2008.90	325	45.87	3.99	4.54	32.35	0.75

$$P_{inj}(t) = \begin{cases} P_0 + a_1 t^{0.25} + a_2 \exp\left(\dfrac{a_3 t}{a_4 + a_5 t}\right), & \text{for } 0 \le t \le t_e \\ P_e, \quad \text{for } t > t_e \end{cases} \tag{3.10}$$

where P_0 is the atmospheric pressure. The parameters that appear in the Eq. (3.10) are represented in Table 3.3. The pre-form was initially considered as having pressure $P_0 = 1013.23$ mbar and temperature $T_0 = 23$ °C.

References

1. Oliveira IR (2014) Infiltration of loaded fluids in porous media via RTM process: theoretical and experimental analyses. Doctoral Thesis, Process in Engineering, Federal University of Campina Grande, Campina Grande, Brazil. (In Portuguese)
2. Oliveira IR, Amico SC, Lima AGB, Lima WMPB (2015) Application of calcium carbonate in resin transfer molding process: an experimental investigation. Materialwiss Werkstofftech 46:24–32
3. Oliveira IR, Amico SC, Luz FF, Barcella R, Bezerra VMF, Lima AGB (2013) Effect of $CaCo_3$ content in resin transfer molding process. Def Diff Forum 334–335:188–192
4. Oliveira IR, Amico SC, Barcella R, Lima AGB (2014) Application of calcium carbonate in resin transfer molding process. Def Diff Forum 353:39–43
5. Rudd CD, Long AC, Kendall KN, Mangin GCE (1997) Liquid moulding technologies: resin transfer moulding, structural reaction injection moulding and related processing techniques. Woodhead Publishing Limited, Cambridge, England

6. Santos MJN, Lima AGB (2017) Manufacturing fiber-reinforced polymer composite using rtm process: an analytical approach. Def Diff Forum 380:60–65
7. Advani SG, Sozer EM (2011) Process modeling in composites manufacturing. CRC Press, New York, USA
8. Levy Neto F, Pardini LC (2006) Structural composites—science and technology, 1st edn. Edgard Blucher, São Paulo, Brazil
9. Luz FF (2011) Comparative analysis of the fluid flow in RTM experiments using commercial applications. Master Thesis, Mines, Metallurgical and Materials Engineering, Federal University of Rio Grande do Sul, Porto Alegre, Brazil. (In Portuguese)
10. Luz FF, Amico SC, Cunha AL, Barbosa ES, Lima AGB (2012) Applying computational analysis in studies of resin transfer molding. Def Diff Forum 326–328:158–163
11. Luz FF, Amico SC, Souza JA, Barbosa ES, Lima AGB (2012) Resin Transfer Molding Process: Fundamentals, numerical computation and experiments. In: Delgado JMPQ, Lima AGB, Silva MV (eds) Numerical analysis of heat and mass transfer in porous media. Springer-Verlag, Heidelberg, Germany
12. Coutinho BG (2013) Mathematical modeling using boundary fitted coordinates and development of a computational simulation for applications in resin transfer molding process. Doctoral Thesis, Process Engineering, Federal University of Campina Grande, Campina Grande, Brazil. (In Portuguese)

Chapter 4
RTM Process Modeling

Abstract This chapter, based on the rigorous theory of fluid flow through porous media (continuum theory), different approaches (analytical) for description of air and resin flow in fibrous media will be presented. In the macroscopic mathematical modeling, both fibrous media and fluids (resin and air) are considered to be incompressible and the effect of resin sorption by fibers has been considered. Herein, by using a relevant and advanced mathematical treatment, different effects of the process parameters (injection pressure, fluid viscosity, porous media permeability, and porosity) and mold geometry in the fluid infiltration process has been analyzed.

4.1 The Physics of the RTM Process

Liquid composite molding process manufactures value-added products and thus, this activity is in constant growth. In present day, new resins and compounds are being developed for different applications [1].

In the RTM process (a technique of the liquid composite molding family), the resin flows through a dry reinforcement (in general fibrous) until the entire mold is filled, occurring the impregnation in the fibers. Independently of the reinforcement shape (preform, fabric or blanket), in an RTM process the system can be considered as a fibrous porous medium or simply a porous medium [2].

Process control plays an important role to the operation of any manufacturing process, and fiber-reinforced polymer composites manufacturing process isn't an exception. The total control of a specific process involves ensuring accuracy and responsiveness in a continuous production environment subject to internal and external disturbances [3].

Then, the total understanding of the phenomena and control of the process parameters are very important for the success of the manufacturing operation.

The porous media permeability is a non-uniform parameter, and thus, anisotropy is verified inside the porous media. The intensity of this effect in dependent on the fiber orientation and fiber architecture to form different fibrous media: prepreg, preforms, mats, among others [4].

J. M. P. Q. Delgado et al., *Transport Phenomena in Liquid Composite Molding Processes*, SpringerBriefs in Applied Sciences and Technology,
https://doi.org/10.1007/978-3-030-12716-9_4

Prepreg is a composite industry's term used for identifies a resin-impregnated fiber, fabric, or mat in flat form, which is stored for later use in hand lay-up or molding operations. Preforms are feedstock for the RTM and SRIM processes, where a reinforcement in the form of a thick two- or three-dimensional fiber architecture is put in the mold cavity and then, resin in injected into the cavity to obtain the composite part [5]. In the fabrication of prepregs, the polymer resin is only partially cured [6].

Further, high fiber volume fraction provokes small interfiber spacing and reduces permeability. Because of the small interfiber spacing during injection process, resin is highly constrained in small volumes between the fibers, which results in stress concentrations and reduced strength in the matrix-dominated directions. In general, for physical situations where high volume fraction is used, lower resin viscosity is recommended to be used during processing.

When the fibrous media is constituted by fiber (especially vegetable fiber), it is possible to find two scales of pores: micropores (small pores) between the fibers and macropores (large pores) between fiber tows (bundles of fibers). Thus, during the resin impregnation process, the pores between the fiber tows are filled at a much earlier time than the smaller pores between the fibers. Then, in the flow front position is find two different regions: a region that contain a saturated flow front and another with an unsaturated flow front. This way, the saturation of the porous media is not instantaneous as in random preforms. Thus, it is possible to occur inter-bundle and intra-bundle void spaces [7].

During polymer composite processing, the resin is polymerized. The reaction can be relatively easy, when epoxy and polyester resins are used, or greatly complicated as is the case for the polyimide systems. Because of this chemical effect (reaction kinetic) resin viscosity is modified, and thus resin flow behavior inside the fibrous media is very different [8].

The rate of cure, or crosslinking, of a thermosetting resin system can be monitored by simply the viscosity of the resin, which increases during this stage. In general, the viscosity of all resins decreases with the increasing temperature, however, for thermoset reins; their temperature is coupled with the cure kinetics, decreasing with degree of cure [9]. Then, during injection process, the resin must quickly fill the heated mold cavity and wet all the individual fibers before much reaction occurs. This strongly reduces large voids and increases the strength of the composite part after processing.

In relation to non-isothermal processing, thermoplastic resin must be heated to a higher temperature than thermoset resin. For any these resins, the processing temperature must be higher than the glass transition temperature, especially for thermoplastic resins [4]. Further, it is very important to notice that, thermoset resin once cured cannot be re-melted or reformed/reshaped.

In some cases, particles are added to resin, in order to increase mechanical properties of the composite and to reduce process costs. This way, during injection process, resin and particles flow inside the fibrous media.

The migration of fine particles in porous media is a great challenge of both academy and industry. The physical problem involves flow behavior, the trapped particles at some pore sites or their migration out of the porous media. The trapped particles inside the porous media reduce permeability, increases fluid mixture viscosity (fluid plus solid particle) and can increases fluid mixture density. This way, the total and comprehensive understanding of the migration of particles can to prevent the occurrence of entrapment or plugging of the porous media, the existence of void spaces into the porous media, to reduce filling time during injection process, and to help in the development of more robust mathematical models.

The adverse effects of plugging normally occur at higher concentration of fine particles, and thus, the total mass flow rate (resin mixed with particles) goes to zero, during injection process. In this case, the pore surface (wettability and roughness) plays as important role in the release and capture of fines, especially in the presence of resin, as occurs in polymer composite manufacturing process. The entrapment can occur because of several parameters such as the pore structure, the size and concentration of fines, and the hydrodynamic and the colloidal conditions of the permeating fluid [10].

As a final comment, from the information given above, it is possible to see that the behind the RTM process can be find the phenomena of heat and mass transfer and fluid flow inside the porous media. These topics will be treated following.

4.2 Mathematical Modeling and Numerical Simulations in RTM Processes: An Overview

The major challenge for modeling and analysis of fiber-reinforced polymer composites manufacturing is related to heterogeneity and anisotropy of these classes of materials. Thus, to model the resin flow in a porous medium (RTM process) is not an easy task. The modeling is complex involving mass and momentum conservations, and Darcy's Law. Sometimes in the situations where heat transfer occurs, the energy conservation equation is used too. The solution of these equations can be obtained analytically (with severe simplifications) and numerically, through commercial software or codes developed by the own user.

The engineer or designer when confronted with a problem has at his disposal three elements of analysis: analytical methods, numerical methods and laboratory experimentation. Analytical methods have many limitations, since they can only be applied to problems that simplifying hypotheses do not deviate the results too far from the real physical phenomenon and in simple geometries. The laboratory experimentation, however, refers to the actual parameters of the process, but can be expensive or in some situations impossible to perform. However, both play an important role in the validation of numerical methods [2]. Finally, the numerical method (computational simulation) practically presents no restrictions, being able to solve complicated problems applied to complex geometries, and to generate results in a relatively short period of time, becoming faster and economical in relation to the other cited methods. The numerical method must solve one or more differential equations, replacing the existing derivatives in the equation with algebraic expressions that involve the unknown variables. Numerical techniques for the resolution of complex engineering and physics problems have been used to obtain satisfactory results, especially due to the development of high-performance computers with large storage capacity [11].

Computer simulations are effective tools that can be helpful to support RTM optimization [12, 13]. In RTM process, computer simulation is used to predict the resin injection behavior and thus to evaluate fill time, as well as to determine the optimum resin injection and exit points. Therefore, some parameters must be known as resin viscosity, reinforcement permeability, pressure gradient and part geometry.

One of the factors that can be emphasized in the study of the RTM process is the determination of the permeability of the fibrous medium which is a measure of resistance to fluid flow (resin) in the fibrous reinforcement. The simple and practical manner to estimation of the permeability towards the flow is comparing the predicted resin flow front position into the mold with experimental data.

Due to its importance, many published books [7, 14, 15], book chapters [16–18] and articles [19–25] are devoted to composite manufacturing applying the RTM techniques. These works report information about experiments and simulations (1D, 2D and 3D analysis) applied to different shapes.

4.3 Fundamental Transport Equations

The dominant transport phenomena during the application of the RTM technique for polymer composite manufacturing are heat transfer and fluid flow due to the cure process and non-isothermal impregnation of the fibrous preform by the resin.

Because of the high costs involved in the RTM process, numerical simulations appear as an excellent alternative to be used for control and optimization of the filling process, and to reduce costs of the process. Then, we need an appropriated mathematical modeling to predict the filling process of the mold. This topic will be well discussed in the next sections.

4.3.1 Fluid Flow

During the mold filling, the resin is forced to flow between the reinforcement (macroscopic flow) and inside the reinforcement (microscopic flow). In this analysis, there are macro-pores between fibers and micro-pores interfiber. Thus, in a macroscopic level, the RTM process is similar to the fluid flow through porous media. Then, to model this physical process we use the Darcy's law (momentum conservation equation) coupled to mass conservation equation, in order, to predict the location of the resin flow front and the fluid pressure distribution as it flows through the fibrous preform.

Following, will be presented an advanced mathematical modeling related to fluid flow through porous media, with particular reference to RTM process, in which exists the fluid absorption phenomena by fibrous preform.

For modeling of the fluid flow through the fibrous preform including fluid sorption by the fiber, we begin with the appropriated mass conservation equation as follows [26-32].

$$\frac{\partial}{\partial t}(\varepsilon\rho) + \nabla \cdot \left(\rho\vec{U}\right) = -S^{M} \tag{4.1}$$

where ε represents the porosity; ρ and $\overrightarrow{U}$ represent the density and mean (superficial) velocity vector of the fluid, respectively, t is the time of the fluid, and S^M represents the source term related to mass flow rate of resin absorbed by the fiber per volume unit.

The simplified momentum equation as applied to fluid flow in porous media well known as generalized Darcy's law. It is given as follows:

$$\overrightarrow{U} = -\frac{k}{\mu}\nabla P' \tag{4.2}$$

where k represents the permeability of the porous media, μ is the resin viscosity, and P' is the modified fluid pressure.

The modified fluid pressure taken into account the contribution of the fluid pressure and gravity effects as follows:

$$P' = P + \rho g z \tag{4.3}$$

where P is the fluid pressure, g is the acceleration due to gravity, and z is the height above a reference point.

Because of the existence of connected void space into the preform, the true (physical) velocity of the fluid into the pores is more than the superficial (mean) velocity. The relationship between these parameters, at any location into the porous medium, is given as follows:

$$\vec{U} = \varepsilon\vec{V} \tag{4.4}$$

where $\overrightarrow{V}$ represents the true fluid velocity vector.

In problems of fluid flow through the fibrous preform including fluid sorption by the fiber, the mass conservation equation is given by Eq. (4.1). However, if the fluid is considered incompressible (constant density), this equation can be simplified and written as follows:

$$\nabla \cdot \overrightarrow{U} = -s \tag{4.5}$$

where $s = S/\rho$, corresponds to the sink term due to delayed saturation of fibrous preform compared to the empty spaces (pores) between the surrounding fibers.

Neglecting the gravitational effects, and substituting the Darcy's Law, Eq. (4.2) into Eq. (4.5), we obtain:

$$\nabla \bullet (-\frac{k}{\mu}\nabla P) = -s \tag{4.6}$$

Now, by considering thermo-physical properties K and μ constants, the Eq. (4.6) takes the form:

$$\nabla^2 P = \frac{s\mu}{k} \tag{4.7}$$

To solve the Eq. (4.7), the following initial and boundary conditions can be used:

(a) $P = P_{inj}$ at the injection point;
(b) $\frac{\partial P}{\partial n} = 0$ on the walls (n is the normal direction to the wall), and
(c) $P = P_{ff}$ at the front line of the fluid or $p = p_{exit}$. In general, p_{ff} and p_{exit} are considered zero (gauge).

4.3.2 *Heat Transfer*

Heat transfer can occur during fluid flow and curing process. Then, this process can be modeled with the energy conservation equation as follows [33]:

$$\frac{\partial}{\partial t}(\rho c_p T) + \nabla \cdot \left(\rho c_p \vec{V} T\right) = \nabla \cdot \left(\hat{k}\nabla T\right) + S^H \tag{4.8}$$

where c_p and $\hat{k}$ represent specific heat and thermal conductivity of the mixture, respectively. These thermal parameters can be given using the mixture rules as follows:

$$\rho = \varepsilon\rho_f + (1 - \varepsilon)\rho_s \tag{4.9}$$

$$c_p = \varepsilon c_{pf} + (1 - \varepsilon)c_{p_s} \tag{4.10}$$

$$\hat{k} = \varepsilon\hat{k}_f + (1 - \varepsilon)\hat{k}_s \tag{4.11}$$

where the subscript f and s represent the fluid (resin) and solid (preform), respectively. The energy equation considers the existence of local thermal equilibrium. This approach states that the resin and fiber temperatures (including other components) are locally equal, at any moment of the resin injection and curing process. Besides, in the curing process the velocity vector $\vec{V}$ is null. Then, we have heat transfer by convection neglected and pure heat conduction phenomenon occurs.

The last term of the energy equation represents the rate of heat generation by chemical reaction, during curing process of the resin. It can be modeled as follows:

$$S^H = H_r \frac{d\alpha}{dt} \tag{4.12}$$

where Hr is the total heat of reaction and α represents the degree of cure.

The degree of cure α of the resin is given as follows [5–8].

$$\alpha = \frac{Q}{H_r} \tag{4.13}$$

where Q is the heat evolved from time $t = 0$ to time t; it is predicted by the kinetic model. In this sense, Lee et al. [34] report information about heat of reaction, degree of cure and viscosity of a specific resin. The simplest kinetic model for describing the cure of reactive resins in given by [1]:

$$\frac{d\alpha}{dt} = \kappa(1 - \alpha)^n \tag{4.14}$$

where α is the conversion term, $\frac{d\alpha}{dt}$ is the rate of reaction, κ is the temperature dependent parameter, and n is the kinetic exponent.

4.4 Isothermal One-Dimensional Flow: An Analytical Approach

4.4.1 Rectilinear Single-Phase Flow

Herein, we are giving emphasis for rectilinear infiltration problem. In this physical situation, the fluid is introduced through an inlet port at the border of the mold. The fluid flow is limited by the parallel wall, and exits the mold through the ventilation points at the other border. Figure 4.1 illustrates the physical problem discussed here [33, 35, 36].

For this simplified physical situation, the following assumptions were considered in the formulation:

(a) At the micro level, the porous medium is constituted of incompressible solid and fluid phases.
(b) The porous medium is fully saturated, with air at the beginning of the process.
(c) The porous medium is considered homogeneous and isotropic.
(d) Thermo-physical properties are invariant with location and time.

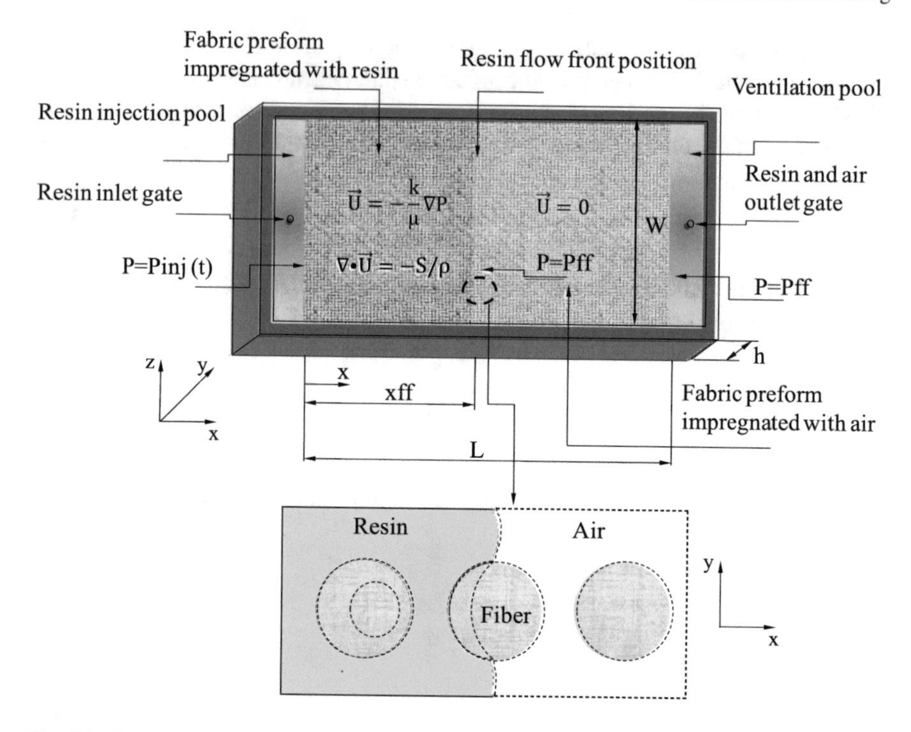

Fig. 4.1 Geometrical configuration of the rectilinear infiltration physical problem

(e) The fluid phases (air and resin) stay separated inside the porous media.
(f) Herein neither a geometrical interpretation of the pore structure nor the exact location of the individual components of the constituents, are considered.
(g) All the phases are in thermodynamic equilibrium.
(h) All the process is isothermal. This means that the energy balance equation is no more necessary.

For a one-dimensional flow of a Newtonian and incompressible fluid through porous media, the Darcy's law can be written as follows:

$$u_x = \frac{Q_x}{A} = -\frac{k}{\mu}\frac{dP}{dx} \tag{4.15}$$

where Qx is the volumetric flow rate, A is the transversal section area of the mold cavity normal to the flow direction, u_x is the superficial velocity or the velocity based on an empty mold cavity, dP/dx is the pressure gradient of the fluid along the reinforcement and x represents the distance in the direction of the forward flow displacement.

In terms of interstitial velocity (u_x) or the flow front velocity (real velocity), we can write the superficial velocity as follows:

$$u_x = \varepsilon v_x = \varepsilon \frac{dx}{dt} \tag{4.16}$$

where ε is the porosity of the fibrous medium, and t is the time.

For a rectilinear and one-dimensional flow, which has fluid velocity equal to zero in the directions y and z, we can write Eq. (4.6) as follows:

$$\frac{d}{dx}\left(\frac{k}{\mu}\frac{dP}{dx}\right) = s \tag{4.17}$$

Then, considering the parameter s constant and integrating the Eq. (4.17) two times, we obtained the following result:

$$P = \frac{s\mu}{k}\frac{x^2}{2} + C_1 x + C_2 \tag{4.18}$$

where C_1 and C_2 are integration constants, that can be obtained by applying the following boundary conditions:

$$x = 0 \Rightarrow P = P_{inj}(t) \tag{4.19}$$

$$x = x_{ff} \Rightarrow P = P_{ff} = 0 \tag{4.20}$$

where P_{inj} is the injection pressure of the resin, x_{ff} and P_{ff} are the position and pressure of the fluid at the flow front, respectively. Then, substituting the Eqs. (4.19) and (4.20) into Eq. (4.18), is obtained:

$$C_1 = -\frac{s\mu}{k}\frac{x_{ff}}{2}\frac{-P_{inj}(t)}{x_{ff}} \tag{4.21}$$

$$C_2 = P_{inj}(t) \tag{4.22}$$

Replacing Eqs. (4.21) and (4.22) into Eq. (4.18), the Eq. (4.23) is obtained, which means that, if the viscosity of the fluid infiltrating an isotropic reinforcement (constant permeability) remains constant, there is a parabolic pressure distribution between the injection point and the flow front.

$$P = \frac{s\mu}{k}\frac{x^2}{2} - \left(\frac{s\mu}{k}\frac{x_{ff}}{2} + \frac{P_{inj}(t)}{x_{ff}}\right)x + P_{inj}(t) \tag{4.23}$$

Derivate of the Eq. (4.23) will be as follows:

$$\frac{dP}{dx} = \frac{s\mu}{k}x - \frac{s\mu}{k}\frac{x_{ff}}{2} - \frac{P_{inj}(t)}{x_{ff}} \tag{4.24}$$

For $x = 0$, we obtain the following pressure gradient:

$$\left.\frac{dP}{dx}\right|_{x=0} = -\frac{s\mu}{k}\frac{x_{ff}}{2} + \left(\frac{s}{2}x_{ff} + \frac{k}{\mu}\frac{P_{inj}(t)}{x_{ff}}\right) \tag{4.25}$$

and for x = 0, we have that:

$$\left.\frac{dP}{dx}\right|_{x=xff} = -sx_{ff} + \left(\frac{s}{2}x_{ff} + \frac{k}{\mu}\frac{P_{inj}(t)}{x_{ff}}\right) \tag{4.26}$$

Then, from the Eq. (4.15) and Fig. 4.1 we can write:

$$u_x(x = 0, t) = \frac{Q(t)}{Wh} = \frac{s}{2}x_{ff} + \frac{k}{\mu}\frac{P_{inj}(t)}{x_{ff}} \tag{4.27}$$

Then, we can determine the injection volumetric flow rate as follows:

$$Q_{inj}(t) = \frac{sWh}{2}x_{ff} + \frac{kWh}{\mu}\frac{P_{inj}(t)}{x_{ff}} \tag{4.28}$$

Using the Eq. (4.16), we can write:

$$u_x(x = x_{ff}) = \varepsilon\frac{dx_{ff}}{dt} = -sx_{ff} + \left(\frac{s}{2}x_{ff} + \frac{k}{\mu}\frac{P_{inj}(t)}{x_{ff}}\right) \tag{4.29}$$

Or yet,

$$x\frac{dx}{dt} = \left[\frac{-s}{2\varepsilon}\right]x^2 + \left[\frac{k}{\mu\varepsilon}P_{inj}(t)\right] \tag{4.30}$$

Then, solving the Eq. (4.30) we obtain:

$$x_{ff}(t) = \sqrt{\frac{2k}{\mu\varepsilon}(e^{\frac{-s}{\varepsilon}t_{ff}}\int_0^{t_{ff}} e^{\frac{s}{\varepsilon}t}P_{inj}(t)dt)} \tag{4.31}$$

By using Eq. (4.31), the time t_{ff} required for a fluid to attain a certain defined position x_{ff} in the mold can be obtained, or vice versa. However, it is necessary to know pressure injection P_{inj} as a function of the filling time.

Substituting the Eqs. (4.29) in the (4.16), we obtain the following equation for the interstitial velocity of the resin flow front:

$$v_x(x = x_{ff}) = \left(\frac{-s}{2\varepsilon}\right)x_{ff} + \frac{k}{\mu\varepsilon}\frac{P_{inj}(t)}{x_{ff}} \tag{4.32}$$

From Eq. (4.32), two possibilities of experimental conditions can be used to keep the rectilinear flow: the injection pressure or injection velocity (in terms of a volumetric flow rate). In this work only the first case will be used.

Table 4.1 Simulation data for rectilinear injection process

Case	V_f (%)	Viscosity (cP)	Maximum injection pressure (bar)	Permeability ($\times 10^{-10}$ m^2)
0% $CaCO_3$	24	330	0.218	3.37

4.4.1.1 Application for Variable Injection Pressure

With the purpose of validation for the mathematical modeling reported before, comparison between predicted and experimental data of the resin flow front position as a function of the process time will be presented. The experimental data used in the simulation are reported in Table 4.1.

Similarly, to the procedure used in the radial injection (item 3.3), collected data of the injection pressure were used in the non-linear regression of the Eq. (4.33) to the experimental data by Santos et al. [33] and Santos and Lima [35]. The cited equation is given below:

$$P_{inj}(t) = \begin{cases} a_1 t^{a_2} + a_3 \exp\left(\dfrac{a_4 t}{a_5 + a_6 t}\right) + a_7, & \text{for } 0 \leq t \leq t_e \\ P_e, \ \text{for } t \;\; t_e \end{cases} \tag{4.33}$$

where P_e is the final pressure (stable pressure) achieved in the experiment and t_e is the instant of time where P is now considered P_e.

Figure 4.2 shows the transient behavior of resin pressure at the injection gate compared to the experimental data reported in the literature [37, 38]. From an analysis of the figure it is noticed that an excellent fit was obtained with a correlation coefficient more than the 0.99 and a variance explained more than the 98% [33, 35]. Table 4.2 summarize the parameters of Eq. (4.33) obtained with the use of the Statistic® software.

It was verified that the pressure, during the first minutes of the injection process increases quickly tending to a steady state condition assuming a value almost constant of $Pe = 21801.8$ Pa ($te = 400$ s). It happens due to the increase in the amount of the resin inside the mold with increasing injection time, which provokes an increase in the fluid flow resistance inside the porous media.

Figure 4.3 illustrates a comparison between simulated and experimental results of the resin front position inside the mold cavity as a function of the injection time. Analyzing this figure, it can be verified that, although the proposed mathematical formulation is simple: one-dimensional and transient formulation, an excellent agreement was obtained between the results, with a local maximum error at the resin front position 10.72% considering a resin absorption term of $s = 0.0001\ s^{-1}$ and an injection time 96 s. This error is a value lesser than the 11.7246% obtained without the consideration of sorption phenomena. It was verified small deviations between the predicted and experimental data which was attributed to the fact that at practice, the flow occurs in the presence of a slightly difference in permeability and viscosity

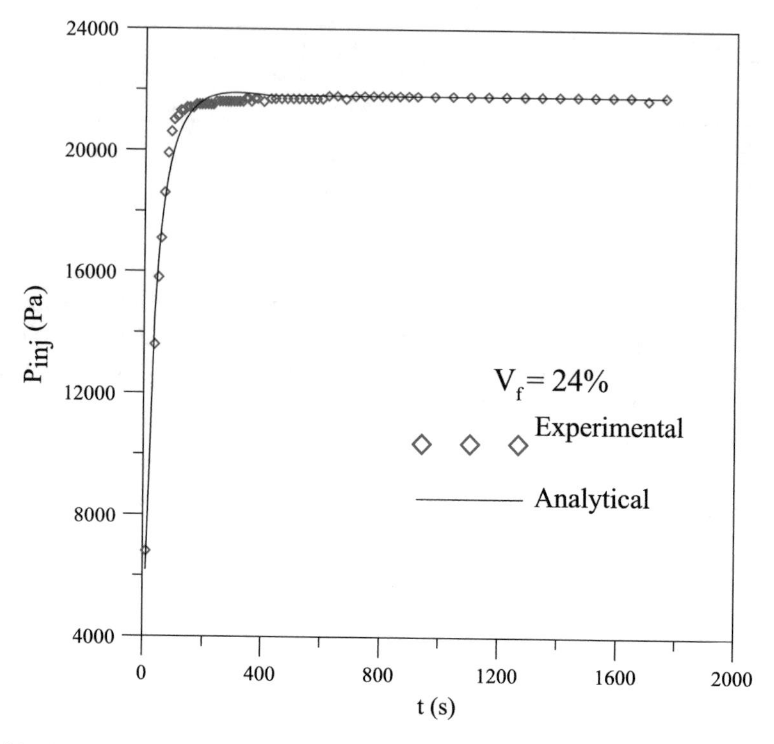

Fig. 4.2 Transient behavior of the resin pressure at the injection gate

Table 4.2 Estimated parameters of Eq. (4.33)

a_1 (bar)	a_2 (−)	a_3 (bar)	a_4 (s^{-1})	a_5 (−)	a_6 (s^{-1})	a_7 (bar)
7.923739	−0.003113	0.003379	42.03634	50.54939	9.168526	−7.87010

from the one at the beginning of the process. It is also noticed an approximately parabolic behavior of the resin flow front position as a function of the injection time, except in the case without to consider the sorption effect (Fig. 4.4). By analyzing the Fig. 4.3, at the end of the process (t = 1758 s), the local error was 0.2469% with the consideration of the sorption phenomena against 5.2778% without the consideration of this phenomena.

Figure 4.5 illustrates the pressure distribution inside the preform as a function of the resin flow front position at different instants of the injection process. From the analysis of this figure, it was verified a parabolic behavior of this parameter at any time of the process. However, it can be observed, at the initial stages of the injection process, higher pressure gradients within the mold. It occurs because two effect: an increase in resin pressure at the injection gate with the time, which tends to stabilize for long times, as shown in Fig. 4.2, and the mass of the

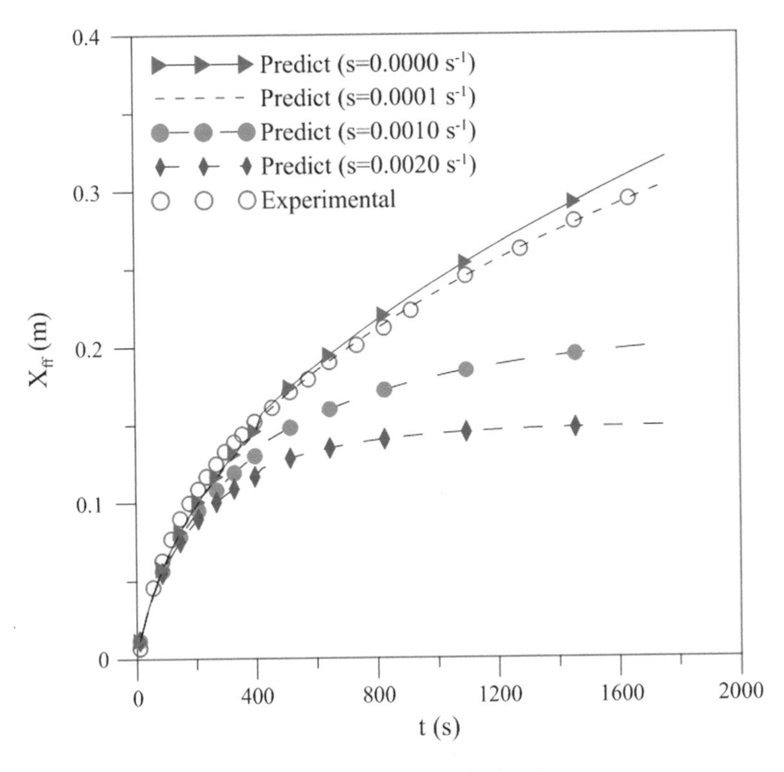

Fig. 4.3 Resin flow front position as a function of the injection time

resin inside the mold, that increases over time. We notice that when the sink term is null, pressure behavior is linear.

Figure 4.6 shows the real/interstitial velocity profile at the resin flow front position and Fig. 4.7 illustrates the resin volumetric flow rate at the inlet gate, both as a function of the process time. From an analysis of these figures is noticed an asymptotic decrease with, both, time of injection, tending to a constant value for long times of process, depending on the sink term value. A more intense sorption effect (higher value of the parameter S) provokes more decay (high value of the velocity temporal variation) in the fluid velocity tending for an approximately null value in shorter injection time. Those instants correspond to the moments when the injection pressure becomes constant and the mold is completely filled by the resin.

4.4.1.2 Application for Constant Injection Pressure

Now, in continuation to this topic, we are presenting simulated results for different physical problems by considering constant injection pressure of the resin at the inlet

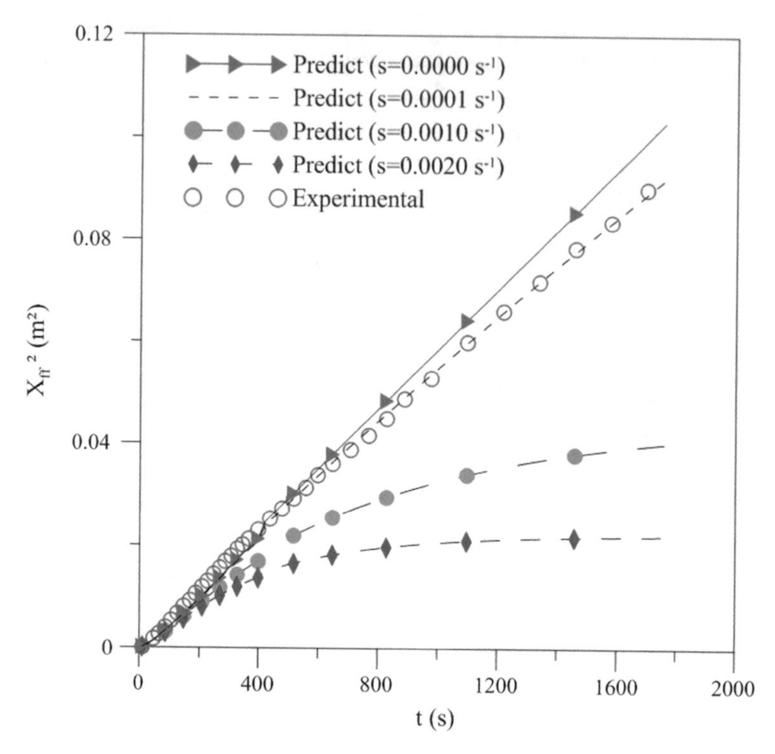

Fig. 4.4 Square of the resin flow front position as a function of the injection time

gate. In Table 4.3 are reported the various cases to be simulated. In all studied cases the sorption term was considered to be 1.0×10^{-4} s^{-1} [36].

Figure 4.8 illustrates the behavior of the resin flow front position and its square, for different values of injection pressure. It is observed that, for the same processing time, the higher the value of P_{inj}, the most distant the flow front position will be from the injection point, evidencing the importance this parameter in the process. The higher the value of P_{inj}, the easily the resin flows through the porous media.

Figure 4.9 illustrates the behavior of the real/interstitial velocity of the resin flow front, and the behavior of the volumetric flow at the injection point of the resin, both as a function of time. From the analysis of this figure we can see an asymptotic decrease in velocity and volumetric flow rate with the injection time, tending to a constant value for long processing times.

Figure 4.10 shows the behavior of the resin front position and its square, as a function of time, for different permeability values of the porous medium. It is observed that, for the same processing time, the higher the value of this parameter, the farthest from the injection point will be the position of the resin flow front. Thus, we state that the higher the value of the permeability, the easier the fluid will flow through the porous medium.

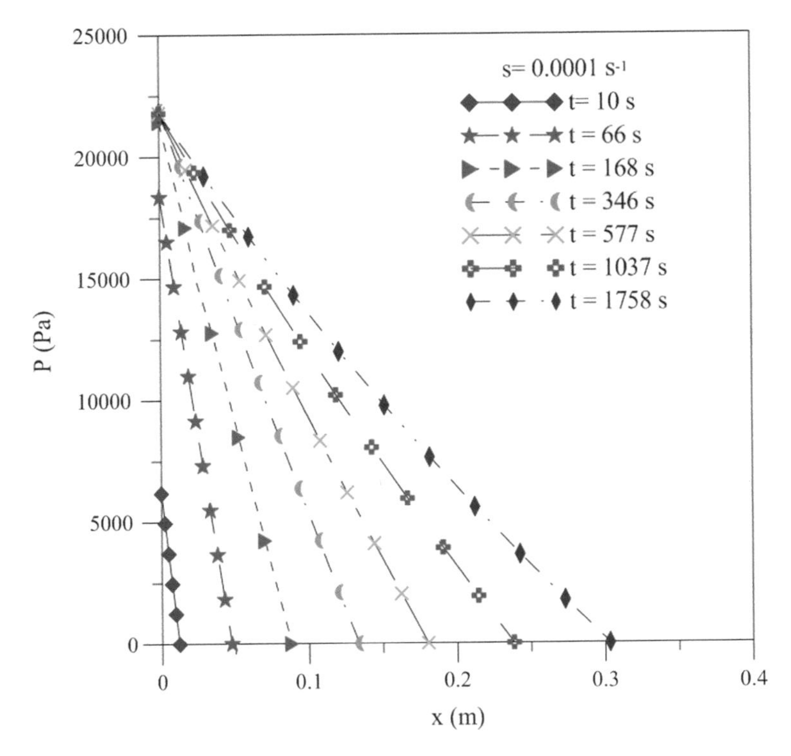

Fig. 4.5 Pressure distribution as a function of the resin front position at different process times

Figure 4.11 illustrates the behavior of the real/interstitial velocity of the resin flow front and resin volumetric flow at the point of injection, both as a function of time, for three values of permeability of the fibrous medium. From the analysis of this figure, an asymptotic decrease in velocity and volumetric flow is observed with the injection time, tending to a constant value for long processing times. It is noticed that the lower the permeability value, the lower the values of these parameters, due to the greater difficulty of the resin to fill the mold.

Figure 4.12 shows the behavior of the resin front position and its square, as a function of time, for different resin viscosity values. It is observed that, for the same processing time, the lower the value of the fluid viscosity, the farthest from the injection point will be the position of the flow front. The smaller the value of this parameter, the greater the ease of the fluid to flow through the porous medium.

Figure 4.13 illustrates the behavior of the real/interstitial velocity of the resin flow front and the resin volumetric flow rate at the injection point, both as a function of time. An asymptotic decrease in velocity and volumetric flow rate is observed, both with the injection time, tending to a constant value for long processing times. It is noted that the lower the resin viscosity value, the higher the values of these parameters, due to the less difficulty of the resin to fill the mold.

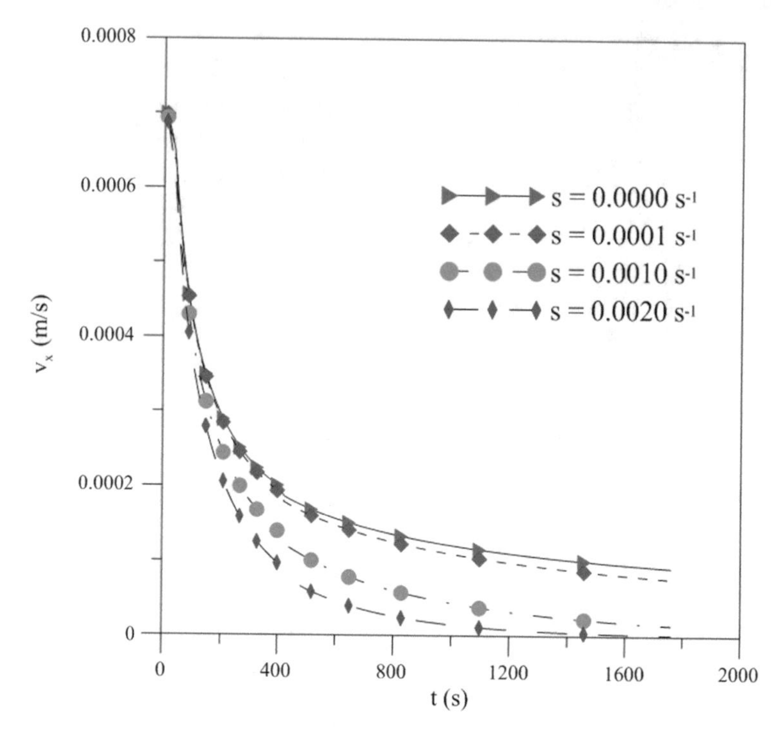

Fig. 4.6 Resin interstitial velocity behavior inside the mold as a function of the injection time

Figure 4.14 shows the behavior of the resin front position and its square, as a function of time, for different values of porosity of the fibrous medium. It is observed that, for the same processing time, the lower the value of this parameter, the farthest from the injection point will be the position of the resin flow front, evidencing the importance of this parameter in the process. The lower the value of the porosity, the greater the fluid flows in the porous medium due to the increase in the real velocity of the resin, fixed the resin injection volumetric flow rate.

Figure 4.15a illustrates the behavior of the real/interstitial velocity of the resin flow front as a function of time. An asymptotic decrease in velocity with the injection time is observed, tending to a constant value for long processing times. It is noticed that, as the porosity value decreases, there is also an increase in the interstitial velocity of the resin, due to the strangulation suffered by the resin when passing through the interstices, causing acceleration in the same. Figure 4.15b illustrates the volumetric injection flow rate of the resin as a function of the injection time within the preform. It is verified that, for higher porosity values, a higher resin injection volumetric flow rate is required to ensure complete mold filling.

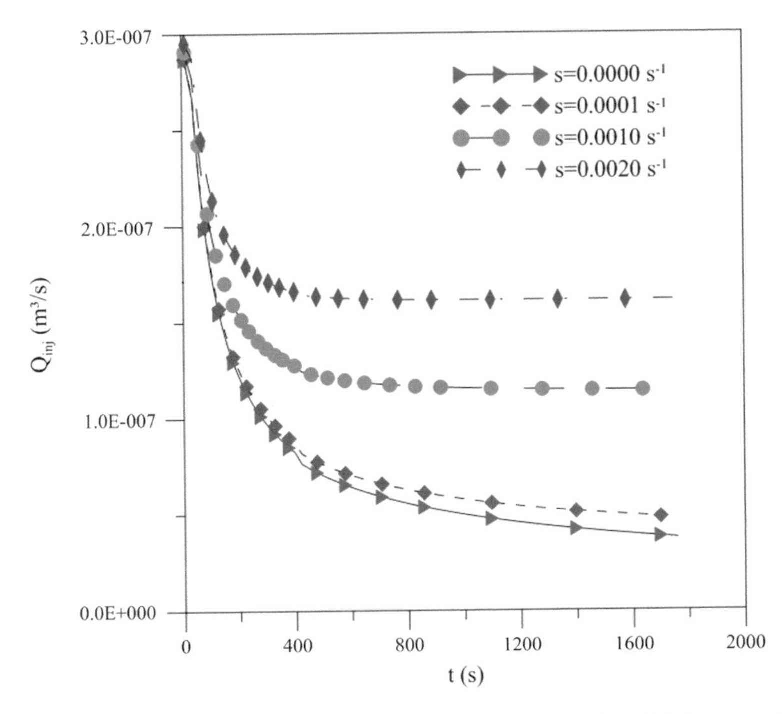

Fig. 4.7 Resin volumetric flow rate behavior at the inlet gate as a function of the injection time

Table 4.3 Simulation data for rectilinear infiltration (constant injection pressure)

Case	Porosity (−)	Viscosity (cP)	Maximum injection pressure (bar)	Permeability ($\times 10^{-10}$ m^2)
1	0.76	330	0.200	3.3700
2	0.76	330	0.218	3.3700
3	0.76	330	0.250	3.3700
4	0.76	330	0.218	0.0337
5	0.76	330	0.218	0.3370
6	0.76	500	0.218	3.3700
7	0.76	1000	0.218	3.3700
8	0.50	330	0.218	3.3700
9	0.60	330	0.218	3.3700

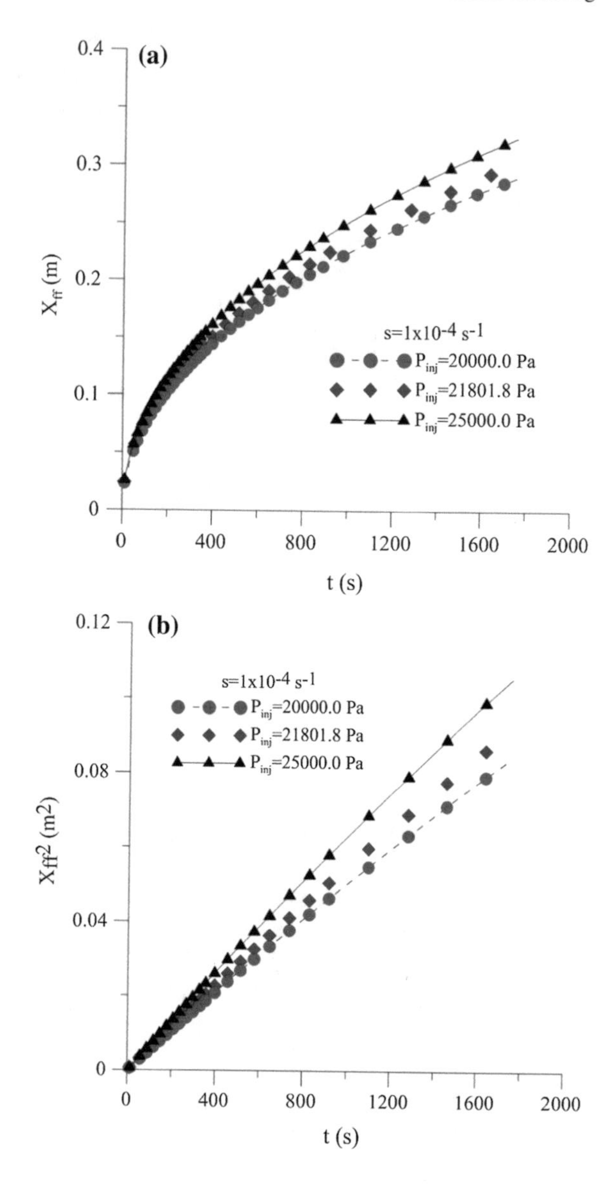

Fig. 4.8 **a** Position and **b** Square of the position of the resin flow front as a function of time for different resin injection pressures

4.4.2 Radial Single-Phase Flow: One-Dimensional Approach

Radial flow is a two-dimensional fluid flow in which the fluid motion is symmetrical around an axis. Then, it presents a one-dimensional flow behavior. This is a simple case of a two-dimensional case.

Figure 4.16 illustrates the radial resin flow in a mold cavity in which exists a fibrous preform. The resin injection point is located at the central region of the mold

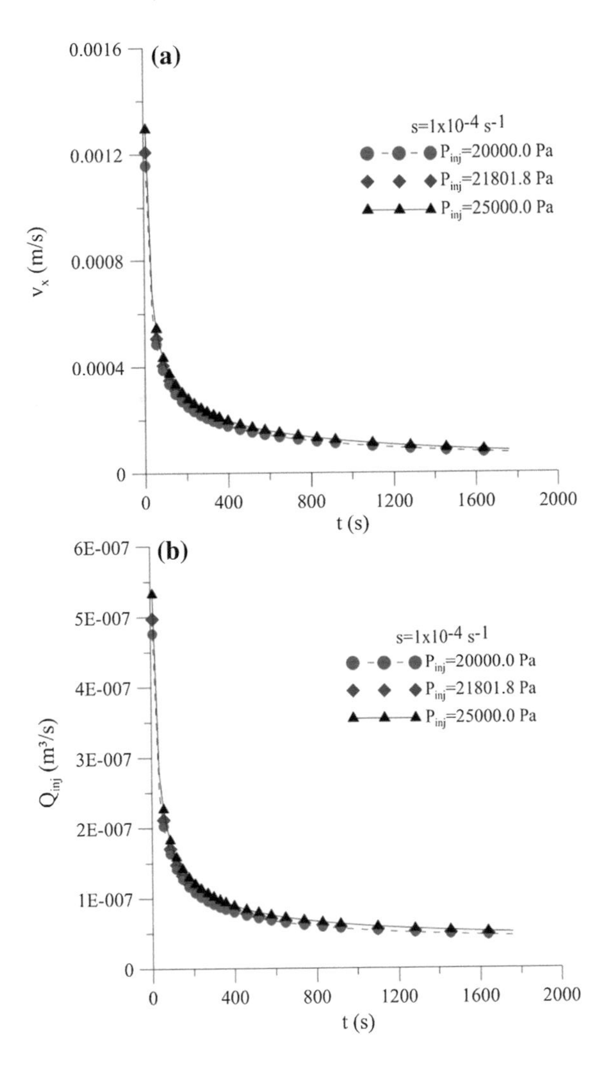

Fig. 4.9 Behavior of the real velocity of the resin flow front and the resin volumetric flow rate of injection as a function of time for different injection pressures

corresponding to a circle of radius r_{inj}, which must not be null. For the modeling of the flow, one must use an appropriate coordinate system, in this case, the cylindrical coordinate system.

Then, the mass conservation equation, without considers the sorption term will be written as follows:

$$\frac{1}{r}\frac{\partial}{\partial r}(r\rho u_r) + \frac{1}{r}\frac{\partial}{\partial \theta}(\rho u_\theta) + \frac{\partial}{\partial z}(\rho u_z) + \frac{\partial \rho}{\partial t} = 0 \tag{4.34}$$

For one-dimensional and incompressible flow, the Eq. (4.34) is simplified as follows:

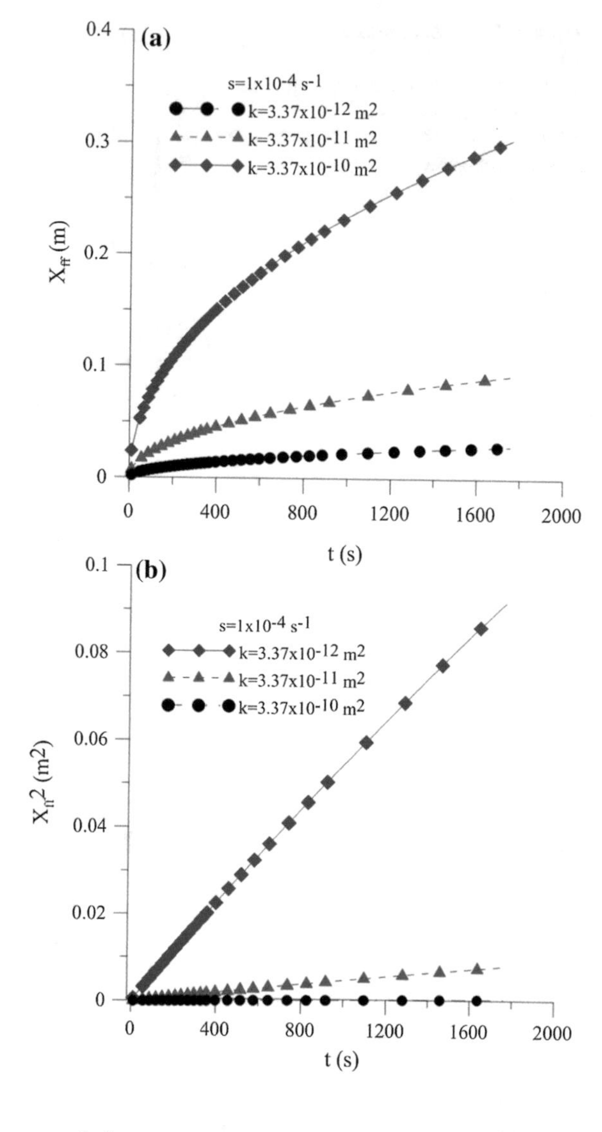

Fig. 4.10 **a** Position and **b** Square of the front position of the resin as a function of the injection time for different permeabilities of the fibrous medium

$$\frac{1}{r}\frac{\partial}{\partial r}(ru_r) = 0 \tag{4.35}$$

On the other hand, Darcy's Law written in cylindrical coordinates, for one-dimensional flow is given by:

$$u_r = -\frac{k}{\mu}\frac{\partial P}{\partial r} \tag{4.36}$$

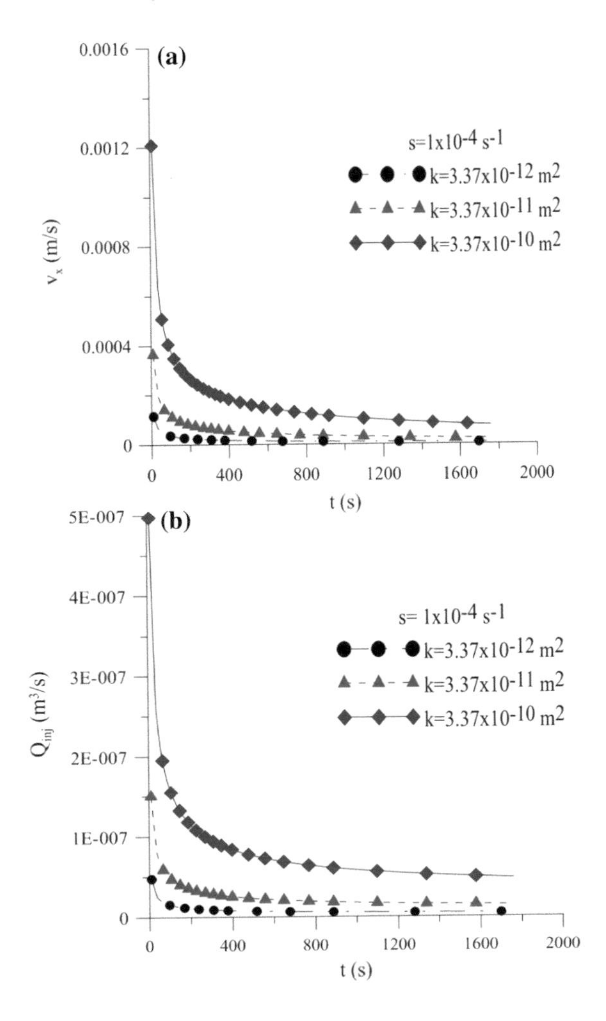

Fig. 4.11 Transient behavior of the resin real velocity at the flow front position and the resin injection volumetric flow rate for different permeabilities of the fibrous medium

Putting Eq. (3.4) into Eq. (4.35), the mass conservation equation will be given as follows:

$$\frac{\partial}{\partial \mathrm{r}}\left(\mathrm{r}\frac{\mathrm{k}}{\mu}\frac{\partial \mathrm{P}}{\partial \mathrm{r}}\right) = 0 \tag{4.37}$$

Integrating Eq. (4.37), the first result is:

$$\mathrm{r}\frac{\mathrm{k}}{\mu}\frac{\mathrm{dP}}{\mathrm{dr}} = \mathrm{C}_1 \tag{4.38}$$

and after,

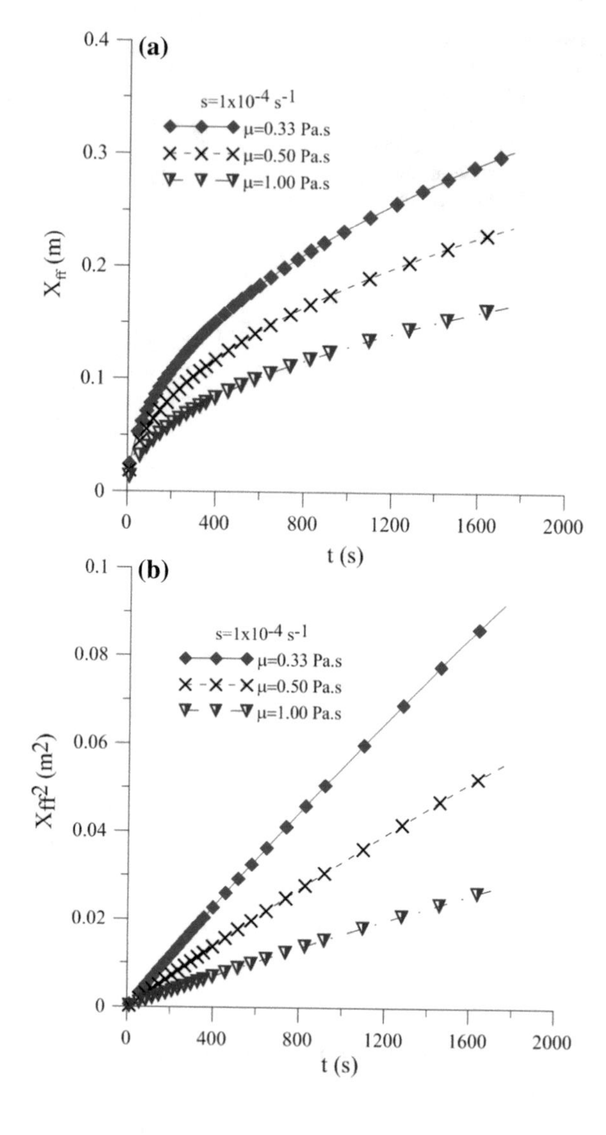

Fig. 4.12 **a** Position and **b** Square of the resin front position as a function of time for different resin viscosities

$$P = C_1 \frac{\mu}{k} \ln r + C_2 \tag{4.39}$$

Equation (4.39) is a general mathematical expression for determine the pressure distribution inside of a radial flow system.

The constants C_1 and C_2 are determined by using the following boundary conditions:

$$r = r_{inj} \Rightarrow P = P_{inj}(t) \tag{4.40}$$

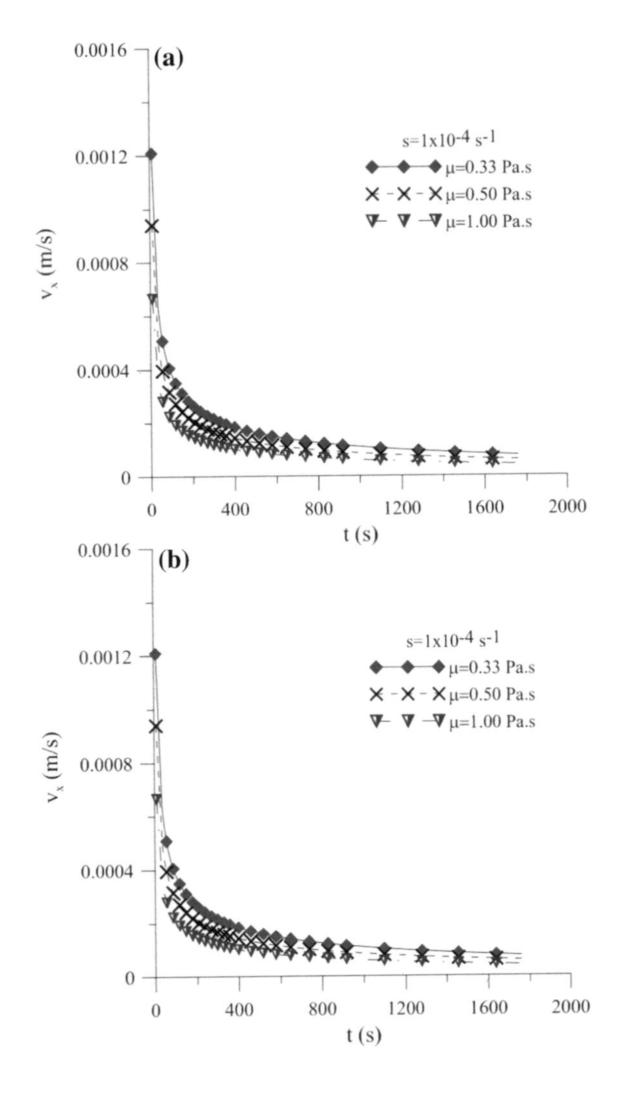

Fig. 4.13 Behavior of the real velocity of the resin front position and the resin injection volumetric flow rate for different resin viscosities

$$r = r_{ff} \Rightarrow P = P_{ff} \tag{4.41}$$

Then, introducing these boundary conditions into the Eq. (4.39), it will result in the following equation:

$$P = \left[P_{inj}(t) - P_{ff}\right] \frac{\ln\left(\frac{r}{r_{ff}}\right)}{\ln\left(\frac{r_{inj}}{r_{ff}}\right)} + P_{ff} \tag{4.42}$$

The derivative of Eq. (4.42) will be given by:

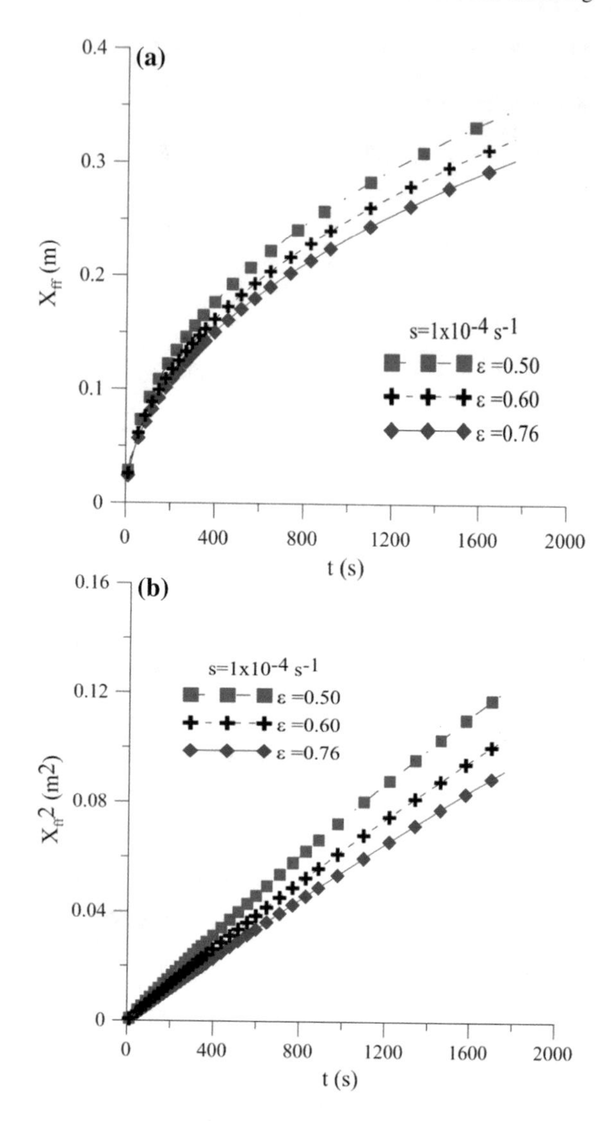

Fig. 4.14 **a** Position and **b** Square of the front position of the resin as a function of the injection time for different porosities of the fibrous medium

$$\frac{dP}{dr} = \frac{\left(P_{inj}(t) - P_{ff}\right)}{\ln\left(\frac{r_{inj}}{r_{ff}}\right)}\left(\frac{1}{r}\right) \tag{4.43}$$

Now, applying those results in the Darcy's Law it will result in:

$$u_r = \frac{k}{r\mu}\frac{\left[P_{inj}(t) - P_{ff}\right]}{\ln\left(\frac{r_{ff}}{r_{inj}}\right)} \tag{4.44}$$

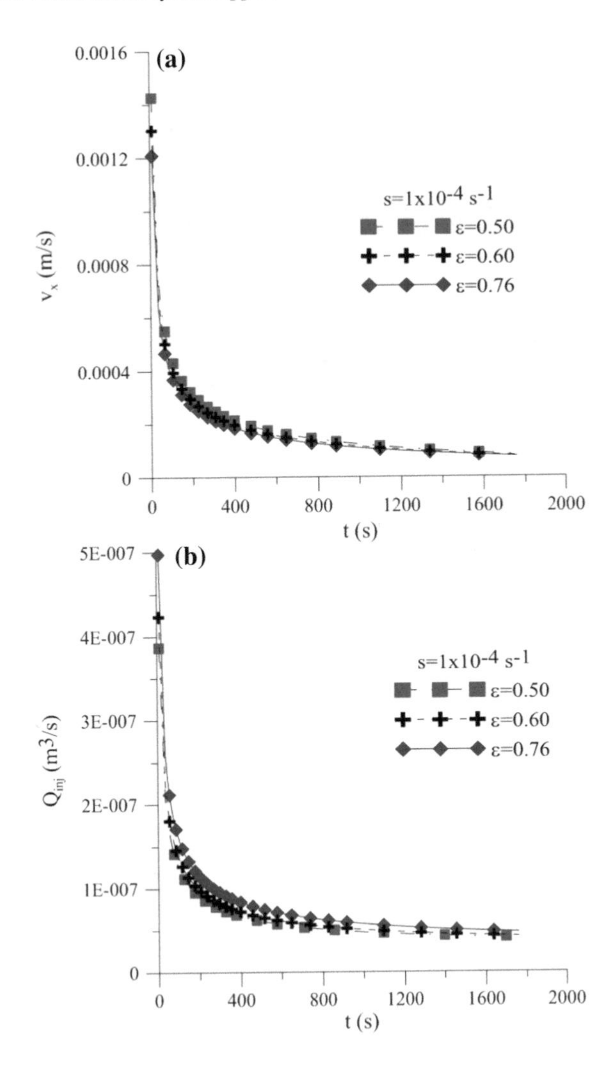

Fig. 4.15 Transient behavior of the real velocity of the resin front position and the injection volumetric flow rate of the resin for different porosities of the fibrous medium

The real velocity will be given as follows:

$$v_r = \frac{u_r}{\varepsilon} = \frac{k}{\varepsilon \mu r} \frac{[P_{inj}(t) - P_{ff}]}{\ln\left(\frac{r_{ff}}{r_{inj}}\right)} \tag{4.45}$$

Then, the injection volumetric flow rate at any time will be given by:

$$Q_{inj} = v_r \cdot A = h \int_0^{2\pi} r v_r d\theta \tag{4.46}$$

or yet,

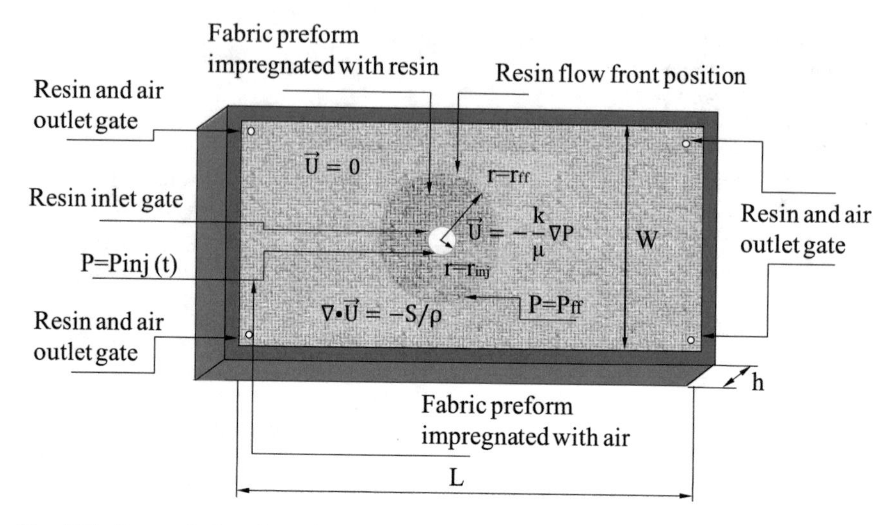

Fig. 4.16 Geometrical configuration of the radial infiltration physical problem

$$Q_{inj} = \frac{k}{\varepsilon \mu r_{inj}} \frac{\left[P_{inj}(t) - P_{ff}\right]}{\ln\left(\frac{r_{ff}}{r_{inj}}\right)} \cdot 2\pi r_{inj} h \tag{4.47}$$

where h represents the thickness of the mold cavity. Further, since that

$$v_r = \frac{u_r}{\varepsilon} \tag{4.48}$$

We can write:

$$\frac{dr}{dt} = \frac{k}{\varepsilon \mu r} \frac{\left[P_{inj}(t) - P_{ff}\right]}{\ln\left(\frac{r_{ff}}{r_{inj}}\right)} \tag{4.49}$$

From the Eq. (4.49), we obtain:

$$r \ln\left(\frac{r}{r_{inj}}\right) dr = \frac{k}{\mu \varepsilon}\left[P_{inj}(t) - P_{ff}\right] dt \tag{4.50}$$

Integrating Eq. (4.50), we obtain the following result at the flow front:

$$\int_{r_{inj}}^{r_{ff}} r \ln\left(\frac{r}{r_{inj}}\right) dr = \int_{0}^{t_{ff}} \frac{k}{\mu \varepsilon}\left[P_{inj}(t) - P_{ff}\right] dt \tag{4.51}$$

After the integration of the left side of the Eq. (4.51), we obtain the following result:

$$\frac{1}{2}r_{ff}^2\ln\left(\frac{r_{ff}}{r_{inj}}\right) - \frac{1}{4}\left[r_{ff}^2 - r_{inj}^2\right] = \frac{k}{\mu\varepsilon}\int_0^{t_{ff}}\left[P_{inj}(t) - P_{ff}\right]dt \quad (4.52)$$

Further, if the injection pressure is considered constant, then, Eq. (4.52) will result in:

$$t_{ff} = \frac{\mu\varepsilon}{2k\left(P_{inj} - P_{ff}\right)}\left[r_{ff}^2\ln\left(\frac{r_{ff}}{r_{inj}}\right) - \frac{1}{2}\left(r_{ff}^2 - r_{inj}^2\right)\right] \quad (4.53)$$

where t_{ff} corresponds to the filling time until the resin reaches specified r_{ff} flow front position. This equation is exactly equal to Eq. (3.9), when P_{ff} is null.

It is important to state that all equations presented in this formulation are valid until the fluid reaches the mold wall. After this point, the flow behavior in a rectangular cavity will not be radial, but a mix of rectilinear and curvilinear (which will gradually tend towards to a rectilinear flow) until the fluid fulfill the mold. By analyzing the Eqs. (4.42) and (4.45), we can note that pressure in a radial flow system varies with the Napierian logarithm of the radial distance from the center of the injection port, and that fluid velocity varies directly with the difference between the injection pressure and the flow front pressure, and inversely with the radial position at any instant of the injection process.

Further, all these parameters vary inversely with the Napierian logarithm of the flow front and injection radius relationship.

References

1. Lee LJ (1997) Liquid composite molding. In: Gutowski TG (ed) Advanced composites manufacturing. Wiley, New York, USA, pp 393–456
2. Luz FF (2011) Comparative analysis of the fluid flow in RTM experiments using commercial applications. Master's Thesis, Mines, Metallurgical and Materials Engineering, Federal University of Rio Grande do Sul, Porto Alegre, Brazil. (In Portuguese)
3. Hardt DE (1997) Process control of thermosetting composites: Context and review. In: Gutowski TG (ed) Advanced composites manufacturing. Wiley, New York, USA, pp 457–486
4. Gutowski TG (1997) A brief introduction to composite materials and manufacturing processes. In: Gutowski TG (ed) Advanced composites manufacturing. Wiley, New York, USA, pp 5–41
5. Mazumdar SK (2002) Composites manufacturing: materials, product and process engineering. CRC Press, Boca Raton, USA
6. Callister WD Jr, Rethwisch DG (2008) Fundamentals of materials science and engineering: an integrated approach, 3rd edn. Wiley, Hoboken, USA
7. Advani SG, Sozer EM (2011) Process modeling in composites manufacturing. CRC Press, New York, USA
8. Kardos JL (1997) The processing science of reactive polymer composites. In: Gutowski TG (ed) Advanced composites manufacturing. Wiley, New York, USA, pp 43–80

9. Bunssel AR, Renard J (2005) Fundamentals of fibre reinforced composite materials. Institute of Physics Publishing, Bristol, UK
10. Khilar KC, Fogler HS (1998) Migration of fines in porous media. Kluwer Academic Publishers, Dordrecht, The Netherlands
11. Gonçalves NDF (2007) Finite volume method in unstructured meshes, Master Thesis, Faculty of Sciences, University of Porto, Porto, Portugal
12. Shojaei A, Ghaffarian SR, Karimian MH (2003) Modeling and simulation approaches in the resin transfer molding process: a review. Polymer Compos 24(4):525–554
13. Long AC (2006) Design and manufacture of textile composites, 1st edn. CRC Press, Boca Raton
14. Rudd CD, Long AC, Kendall KN, Mangin GCE (1997) Liquid moulding technologies: resin transfer moulding, structural reaction injection moulding and related processing techniques. Woodhead Publishing Limited, Cambridge, England
15. Gutowski TG (1997) Advanced composites manufacturing. Wiley, New York, USA
16. Lee LJ (1997) Advanced composites manufacturing, In: Gutowski TG (ed) Liquid composite molding. Wiley, New York, USA
17. Advani SG, Hsião K (2005) Transport phenomena in liquid composites molding processes and their roles in process control and optimization. In: Vafai K (ed) Handbook of porous media. CRC Press, Boca Raton, USA
18. Luz FF, Amico SC, Souza JA, Barbosa ES, Lima AGB (2012) Resin Transfer Molding Process: Fundamentals, numerical computation and experiments In: Delgado JMPQ, Lima AGB, Silva MV (eds) Numerical analysis of heat and mass transfer in porous media. Springer-Verlag, Heidelberg, Germany
19. Chui W, Glimm J, Tangerman F, Jardine A, Madsen J, Donnellan T, Leek R (1995) Porosity migration in RTM. In: Proceedings of the ninth international conference on numerical methods in thermal problems, Atlanta, USA, pp 1323–1334
20. Tan H, Roy T, Pillai KM (2007) Variations in unsaturated flow with flow direction in resin transfer molding: an experimental investigation. Compos Part A Appl Sci Manufac 38(8):1872–1892
21. Oliveira IR, Amico SC, Luz FF, Souza JA, Barcella R, Lima AGB (2013) Resin transfer molding process: a numerical investigation. Def Diff Forum 334:193–198
22. Matsuzaki R, Seto D, Todoroki A, Mizutani Y (2013) In-Situ void content measurements during resin transfer molding. Adv Compos Mater 22(4):239–254
23. Oliveira CP, Souza JA, Isoldi LA, Amico SC (2013) Algebraic rectilinear model for multilayer resin transfer molding injection. J Reinf Plastics Compos 32(1):3–15
24. Robinson MJ, Kosmatka JB (2014) Analysis of the post-filling phase of the vacuum-assisted resin transfer molding process. J Compos Mater 48(13):1547–1559
25. Yang B, Jin T, Li J, Bi F (2014) Simulating the resin flow and stress distributions on mold tools during compression resin transfer molding. J Reinf Plastics Compos. 33(14):1316–1331
26. Harris SD, Ingham DB (2005) Parameter identification within a porous medium using genetic algorithims. In: Vafai K (ed) Handbook of porous media, 2. Edn, Taylor & Francis, Boca Raton, USA, 687–742
27. Nield D, Bejan A (2006) Convection in porous media, 3rd edn. Springer, New York, USA
28. McKibbin R (1998) Mathematical models for heat and mass transport in geothermal systems. In: Ingham DB, Pop I (eds) Transport phenomena in porous media. Oxford, UK, pp 131–154
29. Wang CY (1998) Modeling multiphase flow and transport in porous media. In: Ingham DB, Pop I (eds) Transport phenomena in porous media. Oxford, UK, pp 383–410
30. Bories S, Prat M (2002) Isothermal nucleation and bubble growth in porous media at low supersaturations. In: Ingham DB, Pop I (eds) Transport phenomena in porous media II, Pergamon. The Netherlands, Amsterdam, pp 276–315
31. Baytaş AC, Baytaş AF 92005) Entropy generation in porous media. In: Ingham DB, Pop I (eds) Transport phenomena in porous media III. Elsevier Ltda., Oxford, UK, pp 201–226
32. Ma L, Ingham DB, Pourkashanian MC (2005) Application of fluid flows through porous media in fuel cells. In: Ingham DB, Pop I (eds) Transport phenomena in porous media III. Elsevier Ltda, Oxford, UK, pp 418–440

33. Santos MJN, Delgado JMPQ, Lima AGB, Oliveira IR (2018) Liquid injection molding process in the manufacturing of fibrous composite materials: theory, advanced modeling and engineering applications. In: Delgado JMPQ, Lima AGB (eds) Transport phenomena in multiphase systems, 1st edn. Springer-Verlag, Cham, Switzerland, pp 251–272
34. Lee WI, Loss AC, Springer GS (1982) Heat of reaction, degree of cure and viscosity of Hercules 3501-6 resin. J Compos Mater 16(2):510–520
35. Santos MJN, Lima AGB (2017) Manufacturing fiber-reinforced polymer composite using rtm process: an analytical approach. Def Diff Forum 380:60–65
36. Santos MJN, Delgado JMPQ, Lima AGB, Oliveira IR (2018) Resin flow in porous-fibrous media: an application to polymer composite manufacturing. Diffusion Foundations 20:1–15
37. Oliveira IR (2014) Infiltration of loaded fluids in porous media via RTM process: theoretical and experimental analyses. Doctoral Thesis, Process in Engineering, Federal University of Campina Grande, Campina Grande, Brazil. (In Portuguese)
38. Oliveira IR, Amico SC, Lima AGB, Lima WMPB (2015) Application of calcium carbonate in resin transfer molding process: An experimental investigation. Materialwiss Werkstofftech 46:24–32

Chapter 5
RTM Simulations by CFD

Abstract In this chapter, based on the rigorous theory of fluid flow through porous media and the computational fluid dynamic technique, different numerical approaches for description of air and resin flow in fibrous media will be presented. In the analyses, both fibrous media and fluids (resin and air) are considered to be incompressible. Different 2D and 3D simulations by using CFD are presented and the predicted results of the flow front position are analyzed and compared with experimental data.

5.1 General Comments

Most of the mathematical models for the resin injection in RTM processes consider only one phase, the resin. Such models are quite useful to analyze the macroscopic behavior of the fluid flow and understand some of its important properties. Despite of the use, some limitations of these models can be noticed because air phase within the mold cavity is ignored. Models which are based on two phases (air-resin) take into account the interaction between them, producing better results [1]. For more complex problems, including three-dimensional models with multiple phases, commercial software are generally preferred, but will imply higher costs. For these most complicated cases is necessary to use CFD technique.

Computational fluid dynamics (CFD) is a technique applied to obtain approximate solutions (numerical solutions) of different physical problems involving heat and mass transfer and fluid flow in different geometries. Thus, CFD encompasses a great variety of knowledge at different areas such as Engineering, Physics, Mathematics, Compute science, among others. The application of this fascinating technique has been attractive, because it provides high level of details of the physical problem in study, and becomes easy the analysis of the predicted results. Obviously, the quality of the obtained results depend on the information supplied for the model development and simulators that is to be used. According to the various commercial software that can be used to study the RTM process, we can mention the PAM-RTM of the ESI Group's, the RTM-WORX Polywork and the LIMS of the *University of Delaware*, which are specific applications for RTM, as well as the generic *Computational Fluid*

J. M. P. Q. Delgado et al., *Transport Phenomena in Liquid Composite Molding Processes*, SpringerBriefs in Applied Sciences and Technology,
https://doi.org/10.1007/978-3-030-12716-9_5

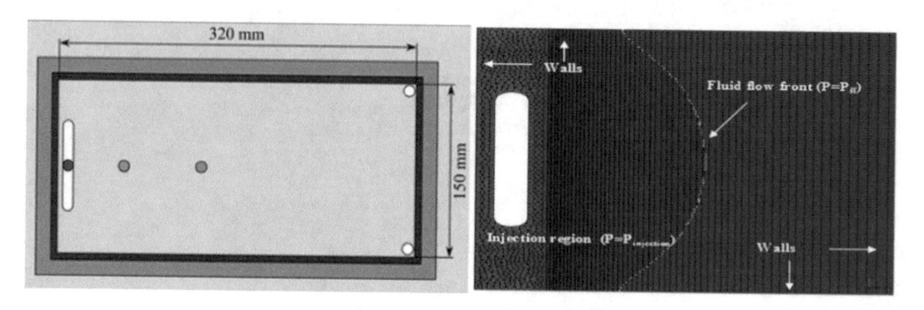

Fig. 5.1 2D mesh used in PAM-RTM computational simulations

Table 5.1 Viscosity and permeability data for each simulation

Case	Viscosity (cP)	Maximum injection pressure (bar)	Permeability ($\times 10^{-3}$ m^2)
0% $CaCO_3$	330	0.203	7.3
40% $CaCO_3$	2113	0.213	12.0

Dynamics (CFD) software such as Ansys CFX and FLUENT, both from Ansys, and Abaqus/CFD (Abaqus), which are simulation tools for fluid mechanics and heat transfer problems, capable of working with complex geometries and simulating the advancement and curing of the resin inside the mold. The PAM-RTM software is used to predict the resin injection profile and, therefore, to evaluate the infiltration time (which should be less than the resin gel time), poor impregnation points and to determine injection and resin/air outlet [2].

Despite of the importance, commercial computational packages also don't allow access to the source code, making difficult or impossible to implement deeper changes to the mathematical and numerical modeling by user.

5.2 Isothermal Two- and Three-Dimensional Flows

5.2.1 Rectilinear Single-Phase Flow

For this purpose, Oliveira [3] and Oliveira et al. [4–7] have realized numerical simulations using the PAM-RTM commercial software.

The mold used in the simulation has dimensions 320 × 150 mm^2 and its geometry (2D mesh) is represented in Fig. 5.1. The mat of this figure has 7077 elements (3685 nodes), with refinement near to the injection region. This mold is shown in Fig. 3.3 too. The experimental data used in the simulation are reported in Table 5.1 and Fig. 3.7 of the item 3.2.

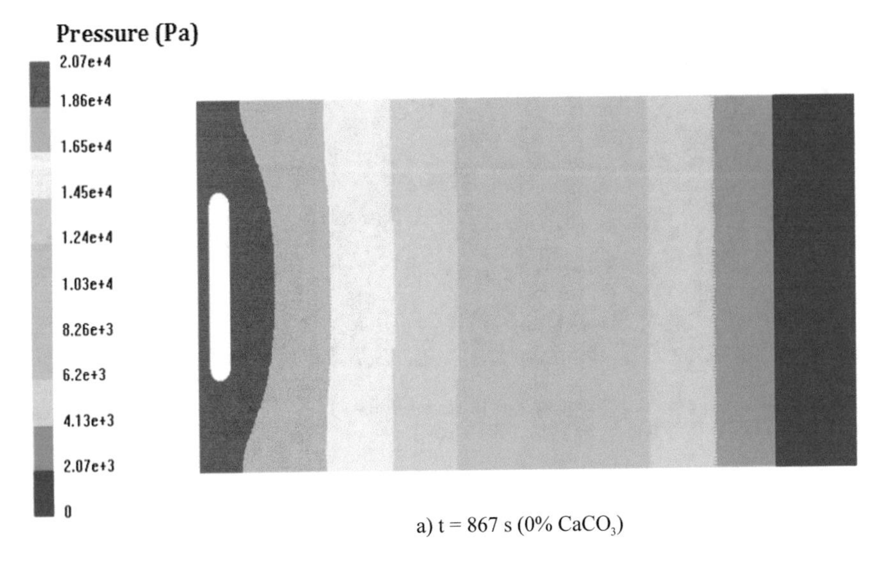

a) t = 867 s (0% $CaCO_3$)

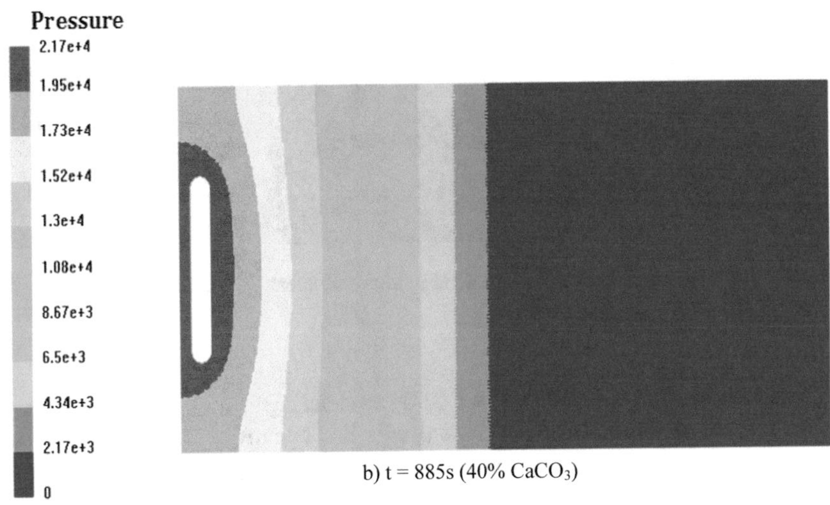

b) t = 885s (40% $CaCO_3$)

Fig. 5.2 Pressure field obtained with PAM-RTM software ($P_{set} = 0.25$ bar, $V_f = 19\%$)

Figure 5.2 illustrates the pressure behavior within the preform for the filling times $t = 867$ s (0% $CaCO_3$) and $t = 885$ s (40% $CaCO_3$). It is perceived that the injection pressure obtained by the PAM-RTM numerical solution approaches the experimental results obtained for both cases as compared with results presented in Fig. 3.7. We can see that the higher pressure occurs in the injection port and lower pressure is verified in the vent port, as expected, because maximum and minimum pressures correspond to boundary conditions for the studied physical problem.

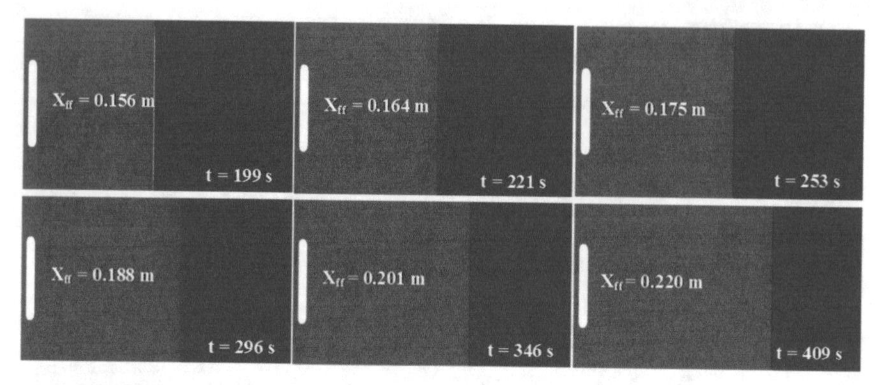

Fig. 5.3 Infiltration profile of the resin inside the mold for glass fiber reinforcement at different filling times ($P_{set} = 0.25$ bar, $V_f = 19\%$, 0% $CaCO_3$)

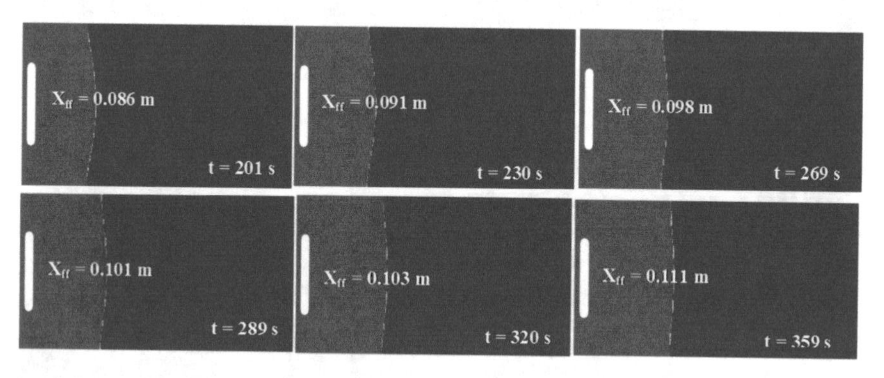

Fig. 5.4 Infiltration profile of the resin for glass fiber reinforcement at different filling times. ($P_{set} = 0.25$ bar, $V_f = 19\%$, 40% $CaCO_3$)

The flow front advance in the 0% $CaCO_3$ case during the resin injection is shown in Fig. 5.3. The rectilinear flow front profile has occurred at approximately half length of the mold. Initially, at the region close to the injection port, the flow has 2D characteristics and the flow front assumes a ring (radial) shape in the main flow direction. When the pressure gradient become linear, the flow front tends to become rectilinear (1D), for example, in $t = 230$ s.

Figure 5.4 shows the predicted resin flow front position along the fibrous media for different processing times for condition 40% $CaCO_3$.

Figures 5.5 and 5.6 show the experimental and numerical flow front position as a function of time for the cases (0% $CaCO_3$ and 40% $CaCO_3$). Results show that the PAM-RTM numerical solution is in good agreement with experimental results. A small error is observed in the nonlinear (close to the injection port) region. The difference is probably due to the 3D characteristic of the experimental setup on which the injection is performed through the bottom of the mold while in the numerical

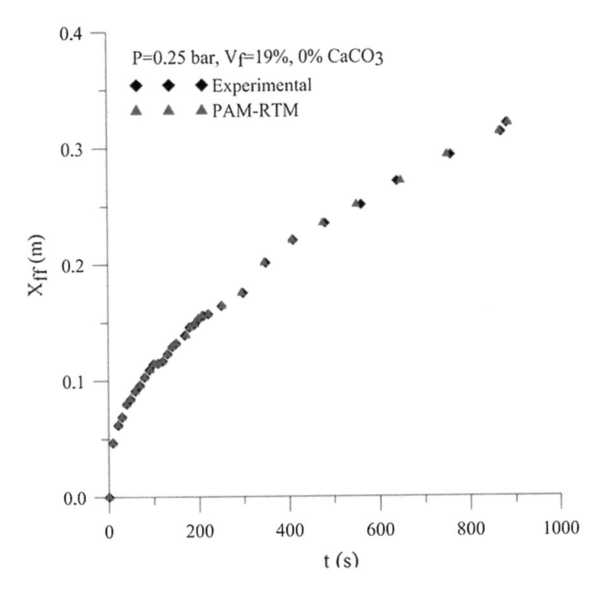

Fig. 5.5 Resin position at the mold versus injection time ($P_{set} = 0.25$ bar, $V_f = 19\%$, 0% $CaCO_3$)

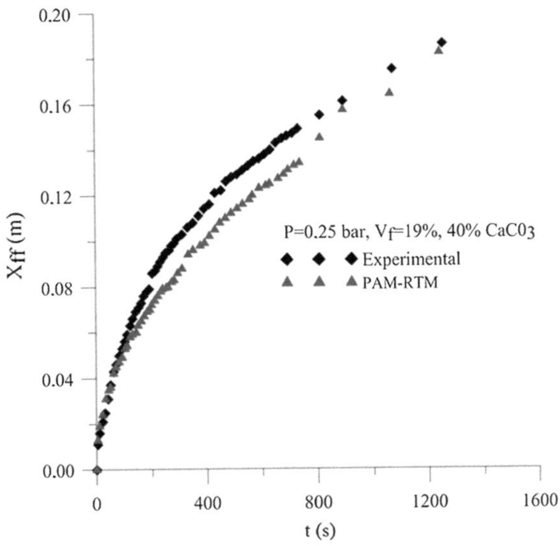

Fig. 5.6 Resin position at the mold versus injection time ($P_{set} = 0.25$ bar, $V_f = 19\%$, 40% $CaCO_3$)

solution prescribed pressure is specified at the borders of the injection hole. Besides, the permeability was determined based or the 1D rectilinear flow (t > 100 s) [2, 8–10].

Comparing the PAM-RTM results with the experimental data for the time 400 s for the filled resin (40% $CaCO_3$) and non-filled resin cases, we can observe the effect of adding Calcium carbonate in the filling time, the higher the $CaCO_3$ concentration in the resin, the bigger the filling time. This is due to the presence of Calcium carbonate

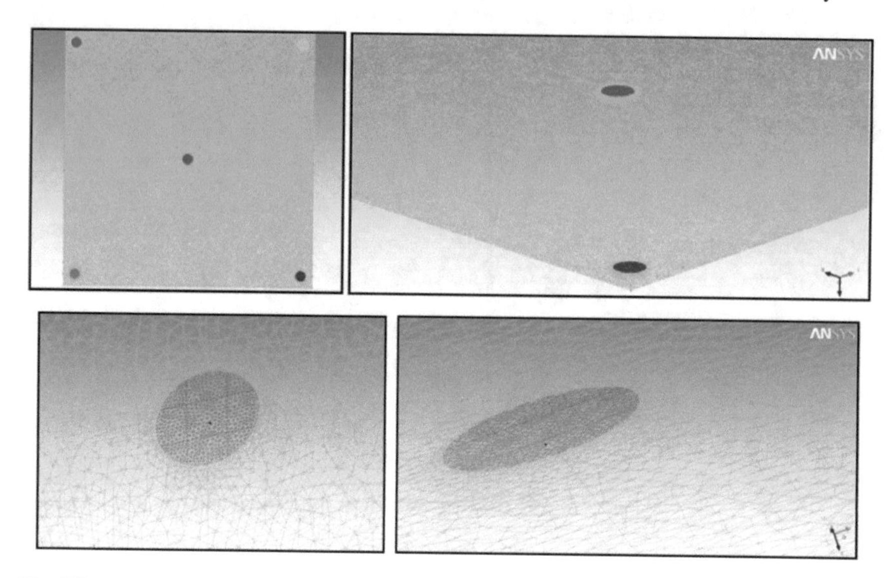

Fig. 5.7 3D mesh used in the Ansys CFX computational simulations

particles among the fibers, hindering the resin flow through the fibrous media. Further, the higher the $CaCO_3$ content in the resin, the higher the fluid mixture viscosity, which affect strongly the filling time.

5.2.2 *Radial Single- and Two-Phase Flows*

Flow simulation in radial system were perfomed by Luz [8] and Luz et al. [11, 12], using the PAM-RTM (2D analysis) and Ansys CFX (3D analysis) commercial softwares.

For 3D simulations, the system is composed by resin, air and solid material (glass fiber). Simulations were performed in a structured grid with 17,532 elements and 23,876 nodes. Figure 5.7 shows the grid used in the study.

In the 2D analysis, the system is composed of resin and solid material (glass fiber). Simulations were performed using a structured grid with 14,819 elements and 16,775 nodes. Figure 5.8 illustrates the grid used in the study. In both 2D and 3D analyses were used simulations data reported in the topic of radial infiltration (Tables 3.2 and 3.3 in the Sect. 3.3).

The volumetric fraction variation of the fluid obtained in the simulations of ANSYS CFX and PAM-RTM were compared and analyzed. Figure 5.9 shows a comparative of the predicted volumetric fraction evolution of the fluid for the case 2 obtained in ANSYS CFX and in PAM-RTM softwares at three different times 60 s, 150 s and 360 s.

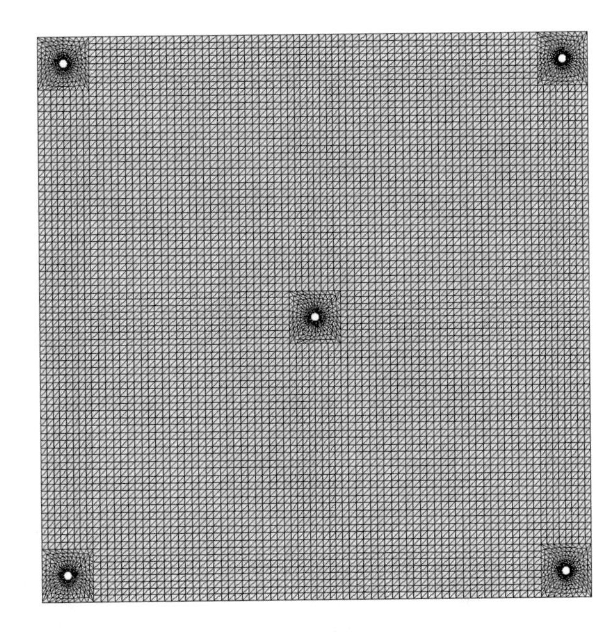

Fig. 5.8 2D mesh used in the PAM-RTM computational simulations

The red (darker) circular areas are regions where the fabric is already impregnated, and the blue (lighter) areas where it is still dry. It is observed that the resin flow front advance radially, however, this impregnated region presents a greater definition, a more regular circle is observed. This is because in the simulations, a homogeneous distribution of the fibers in the fabric is considered, which does not always happen experimentally (see Fig. 3.1). Thus, there is a heterogeneous concentration of fibers along the fibrous reinforcement, thus forming an equally heterogeneous fluid advance.

In Fig. 5.10 is shown a comparative between radial flow front positions, simulated in ANSYS CFX and in PAM-RTM softwares at t_{border} experimental instant for the cases 1, 2 and 3. Experimental results can be seen in Fig. 3.11.

Table 5.2 presents a comparative between experimental t_{border}, simulated in ANSYS CFX and simulated in PAM-RTM, for all cases, showing the respective relative percentual error in relation to the experimental data.

Comparing the different t_{border} it is observed that are not errors superior to 10%, indicating a strong relation between the numerical analysis of the porous media equations and the experimental procedure. Then, it is probable that the errors pointed out in Table 5.2 may be mainly related to the experimental errors, such as lack of homogeneity of the fabric properties, operator error when starting the measurements of time, error in the measurements of the forward radius of flow and error in the measures of time along the process. Further, there are errors related to porous media permeability estimation.

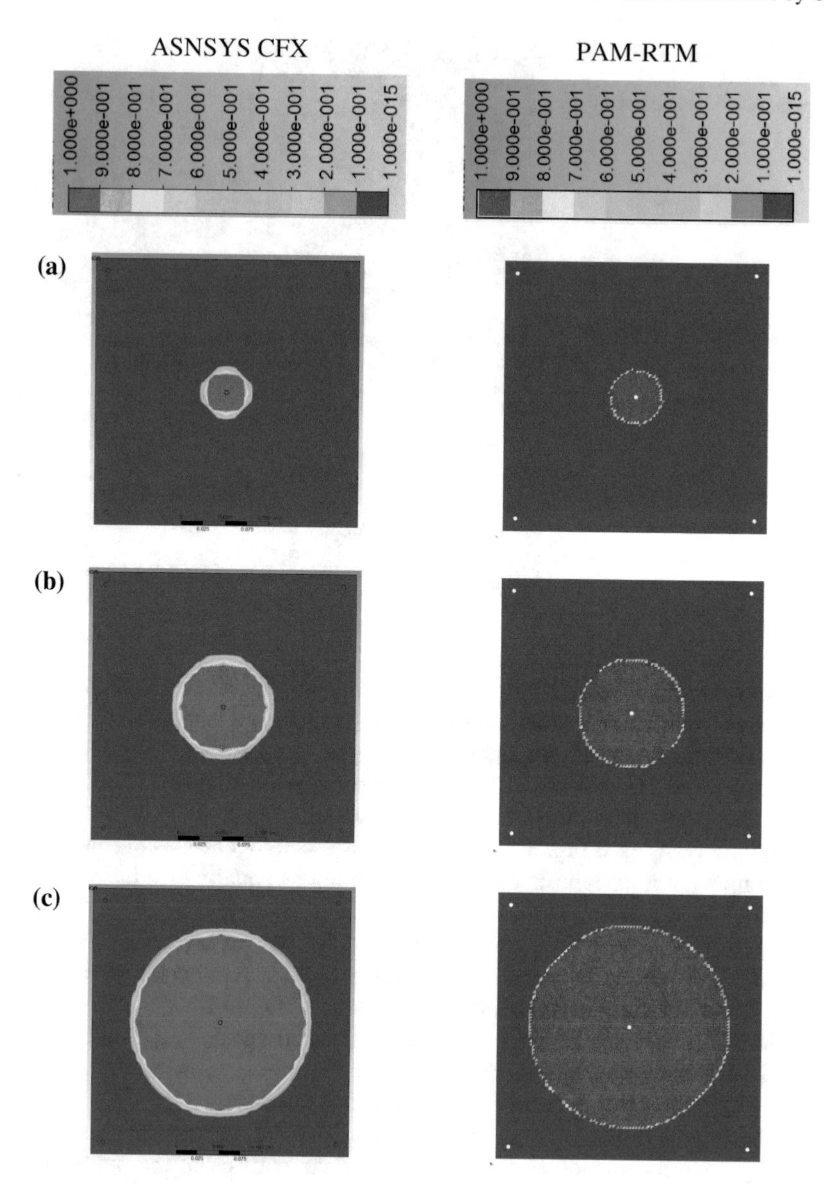

Fig. 5.9 Evolution of the fluid flow front in 60 s (a), 150 s (b) and 360 s (c), for the cases simulated in ANSYS CFX and simulated in PAM-RTM (Case 2)

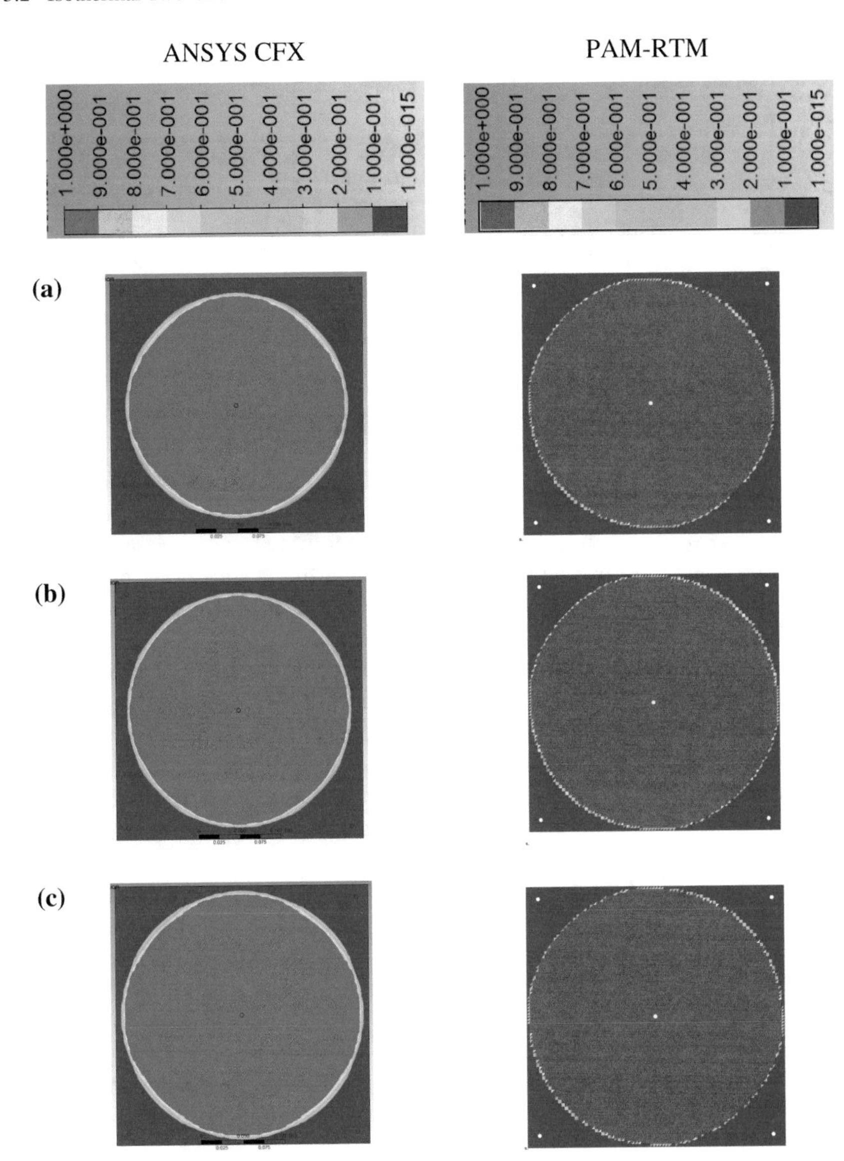

Fig. 5.10 Comparison between resin radial position simulated by ANSYS CFX and PAM-RTM at the experimental t_{border} instant for the cases 1, 2 and 3

Table 5.2 Comparative between the experimental t_{border}, simulated in ANSYS CFX and simulated in PAM-RTM and its respective percentage errors compared to experimental data

Case	t_{border} (s)				
	Experimental (A)	Ansys CFX (B)	Error (%) = (A−B)/A100	PAM-RTM (C)	Error (%) = (A−C)/A × 100
1	640	700	9.38	648	1.25
2	560	588	5.00	568	1.43
3	380	405	6.58	378	−0.53

5.2.3 *Two-Dimensional and Two-Phase Flow: A Boundary-Fitted Curvilinear Coordinate's Application*

A wide variety of practical applications lies in multiphase flow of immiscible fluids in porous media, providing an overview around the models used, highlighting the application of certain techniques [13]. Some works present a review regarding approaches to model and simulate RTM processes [14–16].

In some works, researchers have used discretization methods that need some specific technique for location of the resin front [17–26]. When using the finite element method (FEM) or finite volume based on elements (MEFVC) some authors usually consider the transient problem as a series of quasi static problems, which implies limitations to the magnitude of the time step used [18]. The finite volume method (FVM) is used in fluid flow problems because it ensures the conservation of the physical quantities [27, 28].

Some of the studies conducted to simulate fluid flows during the mold filling stage in RTM processes use one-dimensional approaches (linear or radial injection). In most cases, these investigations are applied to the injection of resin inside rectangular geometries (Cartesian coordinates).

Although the use of Cartesian grids is relatively easy, these grids are limited for the modeling of problems with irregular geometries. In such cases, it is preferable to use boundary fitted coordinates. In this sense, resin injection simulations in RTM processes using the MVF and boundary fitted coordinates are the goal of this topic.

5.2.3.1 Mathematical Modeling

In simulations of RTM processes, the impregnation of the preform is usually modeled as a fluid resin flowing through a homogeneous porous media under isothermal conditions in a macroscopic scale [14, 29]. Here, we use a model for a two-phase (resin–air) immiscible flow with no gravitational and capillarity effects in a homogeneous porous media. The mass conservation equation for any phase p (air and resin) of the system is given as follows [30–32]:

$$\frac{\partial}{\partial t}(\varepsilon\rho^m Z^p) = -\nabla \cdot \left(\rho^p \vec{v}^p\right) - m^p \tag{5.1}$$

where superscript *p* indicates the phase *p*, ε is the porosity of the media, ρ^p, Z^p and $\vec{v}^p$ are, respectively, density, mass fraction and velocity of phase *p*, and ρ^m is the average density of the mixture. In Eq. (5.1), m^p represents the mass flow rate per unit of volume, defined as follows:

$$m^p = \rho^p q^p \tag{5.2}$$

where q^p represents the volumetric flow rate of the phase.

The nature of the resin flow in RTM processes, allows the use of Darcy's law as the momentum conservation equation [1]. For horizontal multiphase systems, the mathematical formulation of the Darcy's law for phase p is written as [33]:

$$\vec{v}^p = -\lambda^p \mathbf{K} \nabla p^p \tag{5.3}$$

where **K** represents the permeability tensor, an intrinsic property of the porous media, λ^p is defined as the phase mobility, given by:

$$\lambda^p = \frac{k^{rel,p}}{\mu^p} \tag{5.4}$$

In Eq. (5.4), μ^p is the phase viscosity and $k^{rel,p}$ is an important parameter that will appear when multiple phases are present in the system. It represents the relative permeability of the phase, which depends directly on its saturation. In the present work, it was assumed: $k^{rel,r} = S^r$ and $k^{rel,a} = 1 - k^{rel,r}$.

Darcy's equation may be combined with mass conservation resulting in the following equation:

$$\frac{\partial}{\partial t}(\varepsilon\rho^m Z^p) = \nabla \cdot \left(\tilde{\lambda}^p K \nabla P^p\right) - m^p \tag{5.5}$$

where $\tilde{\lambda}^p$ is mobility rewritten to include phase density, as follows:

$$\tilde{\lambda}^p = \frac{\rho^p k^{rel,p}}{\mu^p} \tag{5.6}$$

Primary variables in the equations system represented by Eq. (5.5) are the mass fractions (Z^a, Z^r) and pressures (P^a, P^r) of the phases air and resin. Neglecting capillary effects, we have $P^a = P^r = P$. The closing equation needed to solve this system ensures that the entire pore volume is completely filled by phases ($Z^a + Z^r = 1$).

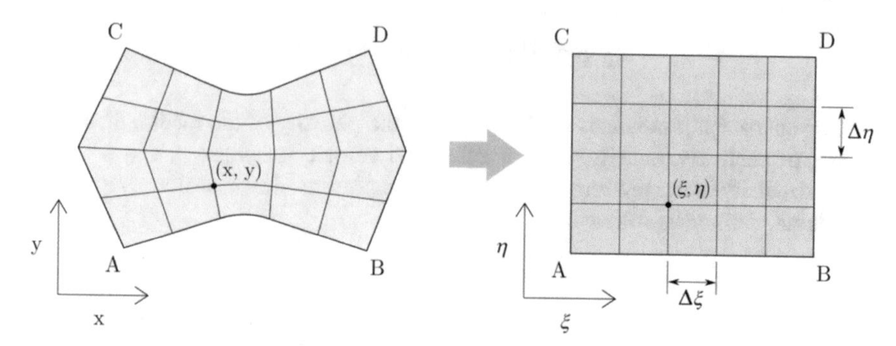

Fig. 5.11 Schematic diagram of the transformation from physical to computational domain

5.2.3.2 Numerical Solution

Boundary-fitted coordinates make mathematical model more flexible to be applied in problems with irregular geometries.

The use of boundary-fitted Coordinates requires that the nonrectangular mesh in the physical space be transformed to a rectangular hypothetical computational representation for it, as pictured in Fig. 5.11.

Governing equations also must be transformed and solved in the computational domain. After transformation, equations can be written as follows:

$$\frac{1}{J}\frac{\partial}{\partial t}(\varepsilon \rho^m Z^p) = \frac{\partial}{\partial \xi}\left(D_1^p\frac{\partial P^p}{\partial \xi} + D_2^p\frac{\partial P^p}{\partial \eta}\right) + \frac{\partial}{\partial \eta}\left(D_2^p\frac{\partial P^p}{\partial \xi} + D_3^p\frac{\partial P^p}{\partial \eta}\right) - \frac{m^p}{J} \quad (5.7)$$

where:

$$D_1^p = \frac{\tilde{\lambda}^p}{J}\left(\xi_x^2 + \xi_y^2\right) \quad (5.8)$$

$$D_2^p = \frac{\tilde{\lambda}^p}{J}(\xi_x\eta_x + \xi_y\eta_y) \quad (5.9)$$

$$D_3^p = \frac{\tilde{\lambda}^p}{J}\left(\eta_x^2 + \eta_y^2\right) \quad (5.10)$$

Equations (5.7)–(5.8) have some terms defined as the metrics of the transformation which bring all the grid information to the transformed equations and relate the physical domain to its computational representation, given by: $\xi_x = Jy_\eta$, $\xi_y = -Jx_\eta$, $\eta_x = -Jy_\xi$ and $\eta_y = Jx_\xi$ [14]. For 2D grids, J represents the relation between areas in computational and physical domains and can be calculated as follows:

$$J = \frac{1}{(x_\xi y_\eta - y_\xi x_\eta)} \quad (5.11)$$

Fig. 5.12 Elementary volume for integration

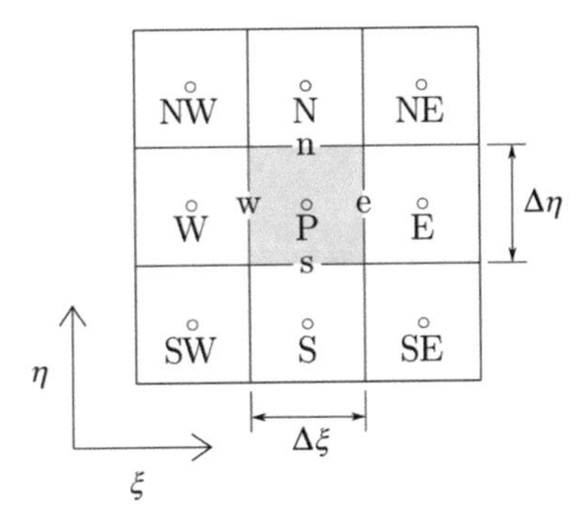

The use of the finite volume method requires the integration of the Eq. (5.8) in space and time for an elementary volume shown on Fig. 5.12. After integration, Eq. (5.8) becomes:

$$\frac{\Delta V}{J}\left[(\varepsilon\rho^m Z^p)_P - (\varepsilon\rho^m Z^p)^o_P\right] = \left[\left(D^p_1\frac{\partial P^p}{\partial\xi} + D^p_2\frac{\partial P^p}{\partial\eta}\right)_e - \left(D^p_1\frac{\partial P^p}{\partial\xi} + D^p_2\frac{\partial P^p}{\partial\eta}\right)_w\right]$$
$$\Delta\eta\Delta\gamma\Delta t + \left[\left(D^p_2\frac{\partial P^p}{\partial\xi} + D^p_3\frac{\partial P^p}{\partial\eta}\right)_n - \left(D^p_2\frac{\partial P^p}{\partial\xi} + D^p_3\frac{\partial P^p}{\partial\eta}\right)_s\right]$$
$$\Delta\xi\Delta\gamma\Delta t - \frac{m^p}{J}\Delta V\Delta t \tag{5.12}$$

where $\Delta V = \Delta\xi\Delta\eta\Delta\gamma$ is the volume dimensions on the boundary-fitted coordinates system.

Some terms with partial derivatives are present in the integrated Eq. (5.12). These terms may be approximated using finite differences, as shows the following samples for the west face, in the Fig. 5.12:

$$\left(\frac{\partial P^p}{\partial\xi}\right)_w = \frac{P^p_P - P^p_W}{\Delta\xi} \tag{5.13}$$

$$\left(\frac{\partial P^p}{\partial\eta}\right)_w = \frac{P^p_N + P^p_{NW} - P^p_S - P^p_{SW}}{4\Delta\eta} \tag{5.14}$$

Phase mobility in Eq. (5.12) must be evaluated in the volume faces (*w*, *e*, *n* and *s*). For this, it was used the Upwind Differencing Scheme. Calculation of the phase mobility in face *e*, as an example, is given by the phase velocity in that face calculated by the following Darcy's equation, rewritten in boundary-fitted coordinates, as follows:

$$\vec{v}^p_e = -\tilde{\lambda}^p_e\left[G_1 e\frac{(P^p_E - P^p_P)}{\Delta\xi} + G_{2e}\frac{(P^p_N + P^p_{NE} - P^p_S - P^p_{SE})}{4\Delta\eta}\right] \tag{5.15}$$

where:

$$G_i = \frac{D_i}{\tilde{\lambda}^p} \quad i = w,\ e,\ n,\ s \tag{5.16}$$

According to Upwind scheme, for $\vec{v}_e^p > 0$, fluid mobility in face *e* will be $\tilde{\lambda}_e^p = \tilde{\lambda}_P^p$. If resin flows in the opposite direction, i.e., $\vec{v}_e^p < 0$, then $\tilde{\lambda}_e^p = \tilde{\lambda}_E^p$. In the other faces, a similar idea can be used.

5.2.4 Initial and Boundary Conditions

The boundary condition for the elementary volumes adjacent to the boundaries is no flow, i.e., it is assumed that all the boundaries are impermeable everywhere inside domain. Thus, the interaction with the surrounding is given by specifying the source/sink terms q^p used in the governing equation. This condition will occur only for volumes with resin inlets or air vents. At these points, it is assumed that the volumetric flow rates of the phases are proportional to their mobility, as follows:

$$\frac{q^r}{\tilde{\lambda}^r} = \frac{q^a}{\tilde{\lambda}^a} = \frac{q^T}{\tilde{\lambda}^T} \tag{5.17}$$

where superscript *T* is the total mobility of the phases, resin and air.

For the resin inlets, it was considered that only resin phase flows. There are two conditions possible for resin injection: constant flow rate or constant pressure. For constant injection flow rate, we have:

$$q^r = q_{def} \tag{5.18}$$

$$q^a = 0 \tag{5.19}$$

For injections under constant pressure, the boundary condition at the resin inlets is given by:

$$P^r = P_{def} \tag{5.20}$$

where subscript <u>def</u> indicates a predefined value.

At the air vents, we can define total volumetric flow rate (resin + air) as the boundary condition as follows:

$$q^T = q_{def} \tag{5.21}$$

$$q^p = \frac{\tilde{\lambda}^p}{\tilde{\lambda}^T} q^T \tag{5.23}$$

Table 5.3 Simulation parameters used in radial infiltration

Parameter	Magnitude	Unit
Maximum injection pressure	149,625	N/m^2
Viscosity	0.0345	Pa.s
Porosity	0.58438	–
Permeability	2.51012×10^{-11}	m^2

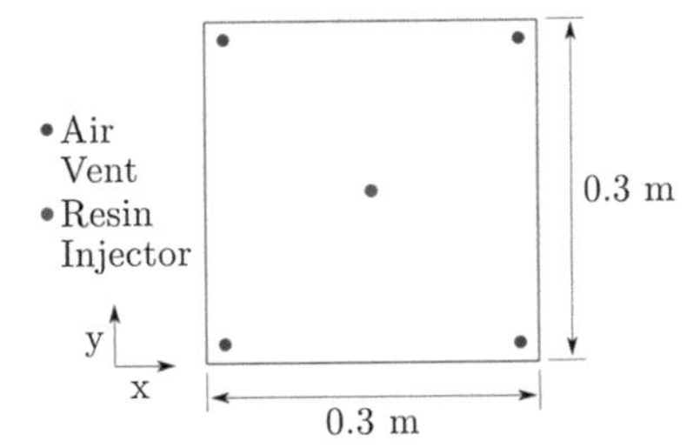

Fig. 5.13 Geometry of the mold cavity for the radial infiltration

Usually, in RTM processes, the condition for the air vent involves the value of the pressure.

$$P^p = P_{def} \tag{5.24}$$

5.2.5 *Applications*

(a) **Regular geometry**: Herein, a case involving radial flow with variable injection pressure was choosen [8, 31]. The process parameters are listed in Table 5.3. Figure 5.13 illustrates the geometry of the square mold which was discretized by a cartesian grid with 75 × 75 volumes. Initially, pressure is atmospheric and increases till reach the constant value. For detail about this physical problem see Sect. 3.3 in this book.

Figure 5.14 shows injection pressure evolution along the time. Figure 5.15 compares digital images captured during experimental tests [8] with predicted results for the flow in three elapsed times: 100, 300 and 600 s. Figure 5.16 shows a comparison between numerical and experimental values for fluid flow front positions along the filling stage time.

As discussed in previous topic, in this book, pressure on filling stage is an important process parameter. Then, some pressure values, inside the mold cavity in positions along horizontal line as illustrated in Fig. 5.17, can be seen in Fig. 5.18 for three filling times. For each time, the higher pressure occurs always at the injection point ($x = 0.15$ m) as expected.

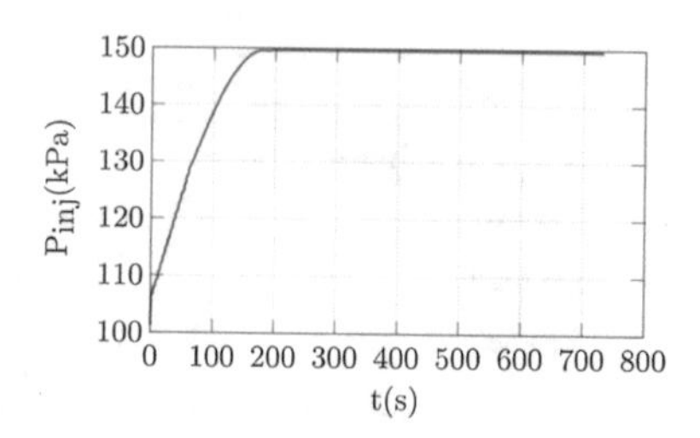

Fig. 5.14 Transient behavior of the injection pressure

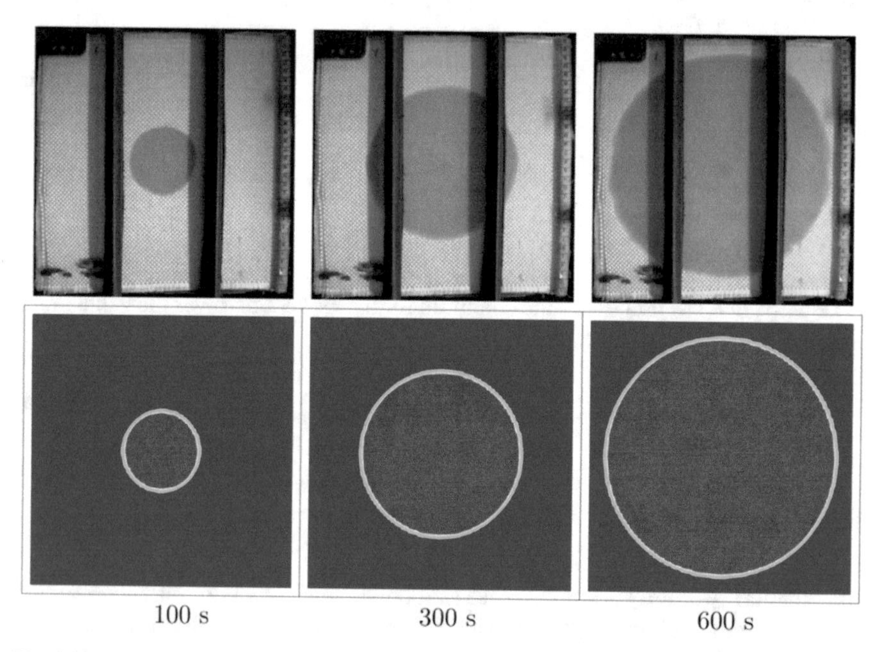

Fig. 5.15 Digital images and numerical results showing resin flow front advance at different times

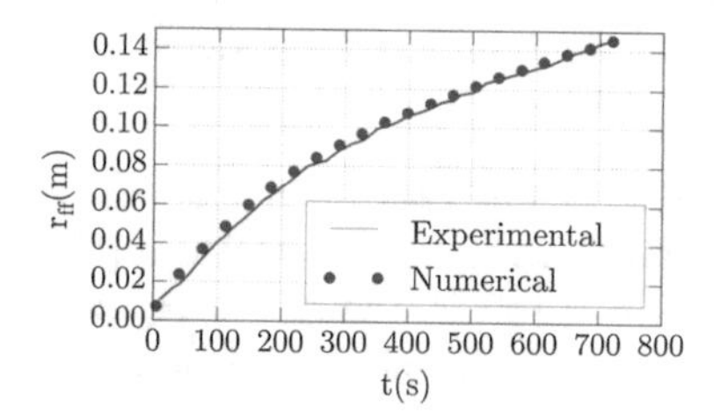

Fig. 5.16 Experimental and predicted fluid flow front radial positions as a function of the filling time

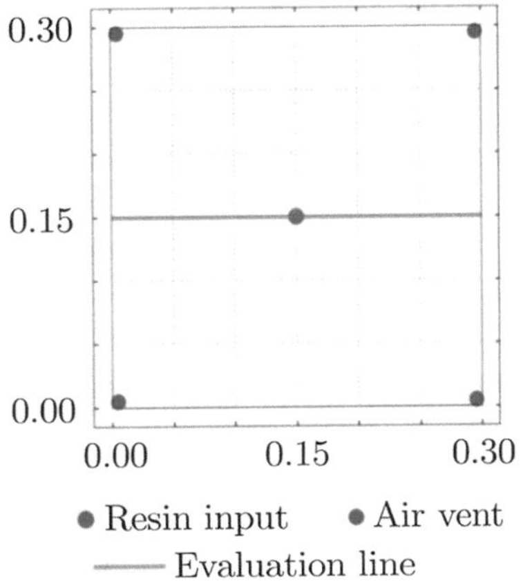

Fig. 5.17 Line used to compute pressure profile inside the mold cavity

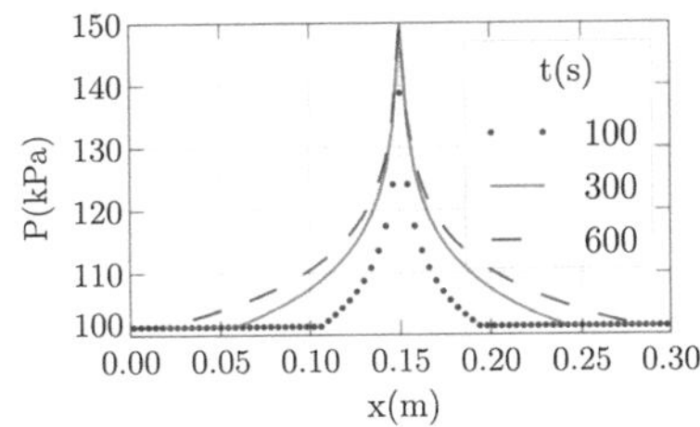

Fig. 5.18 Pressure profile within the mold cavity at different times

From the analysis of the Fig. 5.18 we state that, the initial condition within the mold is: air under atmospheric pressure. At any place, the pressure does not change until the fluid flow front reaches that point. This can be observed analysing pressure curve for instant 100 s. At this moment, the pressure changes from r = 0.1 to r = 0.2 m.

As gauge pressure at the beginning of the process is close to zero, initial flow rate is low and increases until pressure stabilizes 150 s. After this moment, to keep pressure level, volumetric flow rate decreases gradually, as states the Darcy's law. Figure 5.19 shows the transient behavior of the fluid volumetric flow rate at the inlet port.

(b) **Irregular geometry**: For the purpose of studying fluid flow in irregular mold, we can chosen one geometry with irregular borders with 2 mm deep, as shown in Fig. 5.20 [30, 32]. The values of the parameters used in the simulation are listed in Table 5.4. The discretized domain contains 44 $\times$ 35 elementary volumes, as illustrated in Fig. 5.21. In this figure we can see also schematically, the inlets ports and air vents, located at the two opposite sides of the mold cavity. Here, the total volumetric flow rate is divided by the number of control volumes.

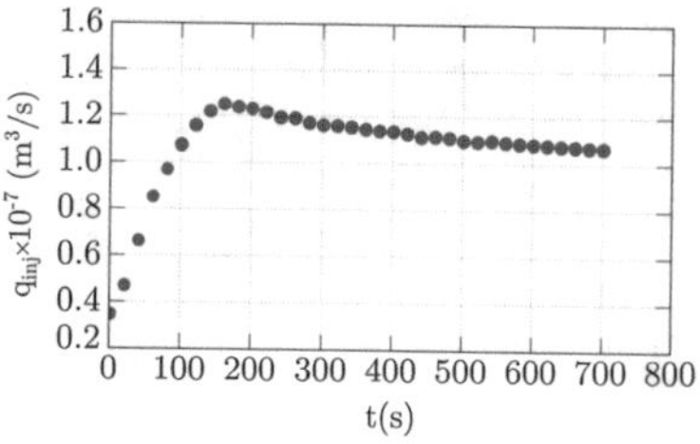

Fig. 5.19 Fluid volumetric flow rate in the injection gate as a function of the time

Table 5.4 Simulation parameters for the irregular geometry case

Parameter	Magnitude	Unit
Injection flow rate, q_{inj}	3.5×10^{-7}	m^3/s
Viscosity	0.1	Pa.s
Porosity	0.7	–
Permeability	2.0×10^{-10}	m^2

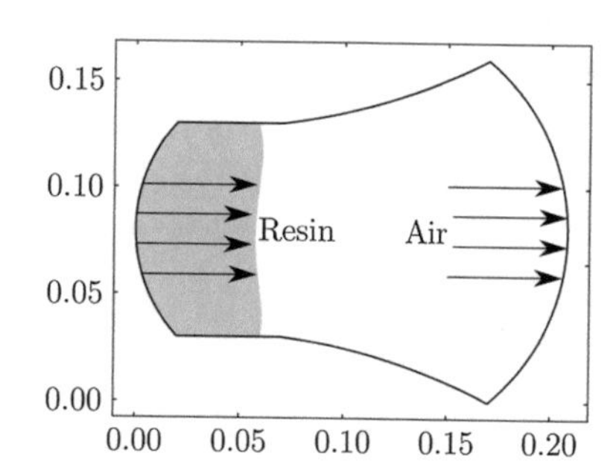

Fig. 5.20 Irregularly-shaped mold used for the multidirectional flow

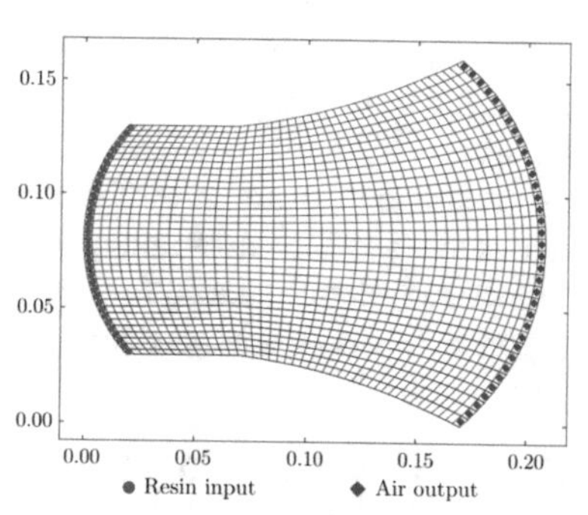

Fig. 5.21 The numerical grid used in the simulations

The evolution of the fluid flow front position may be seen in Fig. 5.22, at three different filling times. As expected, the flow front shape changes according to the border edge. Afterwards, the flow front reaches the end of the cavity unevenly, hitting the corners faster.

The pressure field inside the mold cavity is according to the fluid flow front position, as can be seen in Fig. 5.23 for an elapsed time of 35 s. From the analysis of the figure we can see that the pressure increases from atmospheric level only after

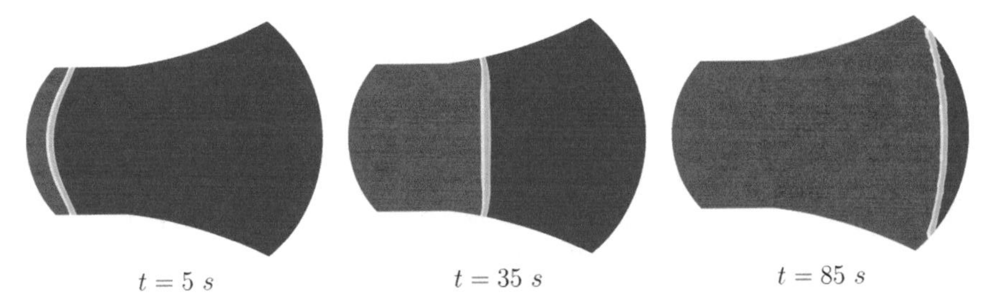

Fig. 5.22 Fluid flow front position at three different times

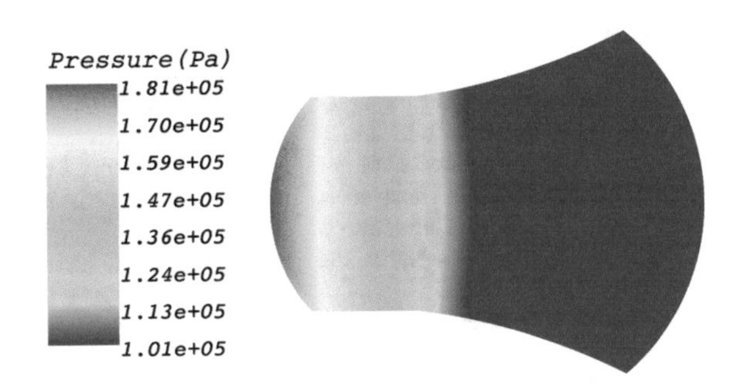

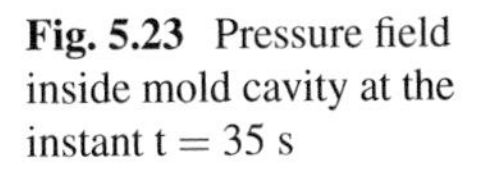

Fig. 5.23 Pressure field inside mold cavity at the instant t = 35 s

resin reaches that area. Note that pressure level is higher at the location where resin is being injected, i.e., along the border, being even higher in the center of that area. As more resin is injected in the cavity, pressure in the center of the mold becomes higher than the border point, and the difference persists until the end of the injection process (Fig. 5.23).

Further, predicted results shows that pressure decreases linearly from the injection points (x = 0 m), where pressure is maximum, until atmospheric value in regions where there is only air. However, we state that this behavior can depend on the local geometry shape of the mold cavity.

References

1. Chui W, Glimm J, Tangerman F, Jardine A, Madsen J, Donnellan T, Leek R (1995) Porosity migration in RTM. In Proceedings of the ninth international conference on numerical methods in thermal problems, Atlanta, USA, pp 1323–1334
2. Tan H, Roy T, Pillai KM (2007) Variations in unsaturated flow with flow direction in resin transfer molding: an experimental investigation. Compos Part A Appl Sci Manufac 38(8):1872–1892
3. Oliveira IR (2014) Infiltration of loaded fluids in porous media via RTM process: theoretical and experimental analyses. Doctoral Thesis, Process in Engineering, Federal University of Campina Grande, Campina Grande, Brazil. (In Portuguese)
4. Oliveira IR, Amico SC, Souza JA, Luz FF, Barcella R, Lima AGB (2013) Resin transfer molding process: a numerical investigation. Def Diff Forum 334–335:193–198

5. Oliveira IR, Amico SC, Souza JA, Lima AGB (2013) Resin transfer molding process: a numerical and experimental investigation. Int J Multiphysics 7(2):125–135
6. Oliveira IR, Amico SC, Souza JA, Lima AGB (2014) Resin transfer molding process: a numerical analysis. Def Diff Forum 353:44–49
7. Oliveira IR, Amico SC, Souza JA, Lima AGB (2015) Numerical analysis of the resin transfer molding process via PAM-RTM software. Def Diff Forum 365:88–93
8. Luz FF (2011) Comparative analysis of the fluid flow in RTM experiments using commercial applications. Master's Thesis, Mines, Metallurgical and Materials Engineering, Federal University of Rio Grande do Sul, Porto Alegre, Brazil. (In Portuguese)
9. Advani SG, Sozer EM (2011) Process modeling in composites manufacturing. CRC Press, New York, USA
10. Lee LJ (1997) Liquid composite molding, In: Gutowski TG (ed), Advanced composites manufacturing. Wiley, New York, USA
11. Luz FF, Amico SC, Cunha AL, Barbosa ES, Lima AGB (2012) Applying computational analysis in studies of resin transfer molding. Def Diff Forum 326–328:158–163
12. Luz FF, Amico SC, Souza JA, Barbosa ES, Lima AGB (2012) Resin transfer molding process: fundamentals, numerical computation and experiments. In: Delgado JMPQ, Lima AGB, Silva MV (eds) Numerical analysis of heat and mass transfer in porous media, 1st edn. Springer-Verlag, Heidelberg Germany, pp 121–151
13. Allen MB III (1985) Numerical modelling of multiphase flow in porous media. Adv Water Resour 8:162–187
14. Shojaei A, Ghaffarian SR, Karimian MH (2003) Modeling and simulation approaches in the resin transfer molding process: a review. Polymer Compos 24(4):525–554
15. Shojaei A (2006) Numerical simulation of three-dimensional flow and analysis of filling process in compression resin transfer moulding. Compos Part A Appl Sci Manufac 37(9):1434–1450
16. Chui WK, Glimm J, Tangerman FM, Jardine AP, Madsen JS, Donnellan TM, Leek R (1997) Process modeling in resin transfer molding as a method to enhance product quality. SIAM Re pp 714–727
17. Phelan FR Jr (1997) Simulation of the injection process in resin transfer molding. Polymer Compos. 18(4):460–476
18. Shojaei A, Ghaffarian SR, Karimian SMH (2003) Numerical simulation of three-dimensional mold filling in resin transfer molding. J Reinf Plastics Compos 22(16):1497–1529
19. Lee SH, Yang M, Song YS, Kim SY, Youn JR (2009) Three-dimensional flow simulation of resin transfer molding utilizing multilayered fiber preform. J Appl Polymer Sci 114(3):1803–1812
20. Shi F, Dong X (2011) 3D numerical simulation of filling and curing processes in non-isothermal RTM process cycle. Finite Elem Anal Des 47(7):764–770
21. Dai F, Zhang B, Du S (2004) Analysis of upper- and lower- limits of fill time in resin transfer mold filling simulation. J Compos Mater 38(13):1115–1136
22. Chang R-Y, Yang W-H (2001) Numerical simulation of mold filling in injection molding using a three-dimensional finite volume approach. Inter J Num Method Fluids 37(2):125–148
23. Yang J, Jia Y, Sun S, Ma D, Shi T, An L (2006) Mesoscopic simulation of the impregnating process of unidirectional fibrous preform in resin transfer molding. Mater Sci Eng A, 435–436:515–520
24. Schmidt FM, Lafleur P, Berthet F, Devos P (1999) Numerical simulation of resin transfer molding using linear Boundary Element Method. Polymer Compos 20(6):725–732
25. Soukane S, Trochu F (2006) Application of the level set method to the simulation of resin transfer molding. Compos Sci Technol 66(7–8):1067–1080
26. Gantois R, Cantarel A, Felices J-N, Schmidt F (2010) Numerical simulation of resin transfer molding using BEM and Level Set Method. Inter J Mater Forming 3:635–638
27. Maliska CR (2004) Computational fluid mechanics and heat transfer, 2nd edn. LTC, Rio de Janeiro, Brazil (In Portuguese)
28. Malalasekera W, Versteeg HK (1995) An introduction to computational fluid dynamics: the finite volume method. Pearson Education Limited, Harlow, England
29. Vafai K (2005) Handbook of porous media, 2nd edn. CRC Press, Boca Raton, USA

30. Coutinho BG (2013) Mathematical modeling using boundary fitted coordinates and development of a computational simulator for applications in resin transfer molding processes. Thesis. Doctoral Thesis, Process Engineering, Federal University of Campina Grande, Campina Grande, Brazil. (In Portuguese)
31. Coutinho BG, Bezerra VMF, Farias Neto SR, Lima AGB (2014) Modelling and simulation of resin transfer moulding: a finite volume approach applied to the mold filling stage. Def Diff Forum 353:67–72
32. Coutinho BG, Bezerra VMF, Farias Neto SR, Lima AGB (2014) A mathematical model for RTM process simulations in geometries with irregular shapes. Int J Multiphysics 8(3):285–296
33. Kumar V (2012) Coupling of free flow and flow in porous media—Dimensional analysis and numerical investigation. Diplomarbeit, Institut für Wasserbau, Universität Stuttgart, Stuttgart, Germany

Chapter 6
Conclusions

Abstract This book is entirely dedicated to polymer composites. In them, an overview of the basic issues related to this class of material ranging from foundations, advanced modeling to engineering applications is given.

Composite can be defined as a multiphase material consisting of two or more distinct materials (micro and macro constituents) which differs in chemical forms and composition, and that are insoluble in each other. It is composed basically of one continuous phase (matrix) and one or more dispersed phases (reinforcement).

A very wide range of possible reinforcement exist which can be used to reinforce all classes of composites. Among them, an important component is the fiber. Fiber reinforcement come in different forms (linear and interlaced fiber structure), and the glass and carbon fibers are the most used.

The resin used for composite materials include thermosetting or thermoplastic resins. The majority of the composites use thermoset resin like matrix, with particular attention to epoxies and polyesters.

There are several polymer composites manufacturing techniques such as, spray-up, hand lay-up, filament winding, pultrusion, vacuum bag molding, autoclave molding, compression molding, liquid composite molding, among others. In this context, the choice of an appropriated manufacture technique and process control are essential to obtain product with good quality (finishing and structural performance) and low cost process.

Polymer composites have expanded their use in different sectors of modern industry, such as automotive, aerospace, sports, civil construction, naval, military and biomedical.

The above applications are given only as examples of the great variety of composite products.

The content of this book has demonstrated that the resin transfer molding technique (variety of liquid molding process) is now a well-understood topic. The information related in this book proved that the RTM process is an efficient technique in the manufacturing of fiber-reinforced polymer composite. This information is related to the various advantages presented by this manufacturing method.

J. M. P. Q. Delgado et al., *Transport Phenomena in Liquid Composite Molding Processes*, SpringerBriefs in Applied Sciences and Technology,
https://doi.org/10.1007/978-3-030-12716-9_6

An advantage of the RTM technique is that the final shape of the composite can often be achieved in one operation with little need for further expensive finishing process.

The mechanisms governing composite behavior in operations are very different from those which are encountered with traditional materials, and requires a full understanding of their constituents (reinforcing and matrix) and their interaction at the fiber-matrix interface (adhesion) or interphase. The adhesion quality is vital to mechanical performance of the composite, in order to avoid failure and damage in the composite (cracks, local deterioration, delamination).

Then, we have a lack of studying the polymer composite manufacturing, specially resin transfer molding process.

In this sense, an understanding of the resin flow front position during injection process is vital to control the process and thus, to obtain good quality of the composite after processing.

For this, standard laboratory experiments were carried out, in order to verify the flow front displacement and to estimate the porous media permeability. Two physical situations were studied: rectilinear and radial infiltration.

Further, in order to describe the fluid motions through porous media, the problem was formulated at a scale at which standard continuum theory applied to immiscible flow in porous media can be used. The advanced mathematical model predicts adequately the two-phase flow behavior inside the fibrous media and in specific cases, the interaction between the materials (air, resin and fibers).

It was showed that all rigorous mathematical treatment given to obtain the analytical and numerical solutions of the governing equations have been useful in describing the fluid flow inside the porous media. Thus, the numerical simulations proved to be an essential tool in the understanding of the RTM process.

From the presented results was verified that, both the numerical and experimental analyses of composite materials manufacture have showed that are crucial in the design of parts and that can be used to help designers, engineers and academics to make design decisions.

It was observed that $CaCO_3$ content mixed with resin, injection pressure and gate location to inlet and outlet resin affect strongly the filling time and void formation in the product. The study shows that the consideration of the sorption term in the mass conservation equation proved to be a more complete mathematical formulation to be used in the RTM process with great success. This term affects strongly the resin flow behavior inside the porous media.

Further, because the great variety of shape of parts, the use of boundary fitted coordinates allowed application of the model to predict resin flow through fibrous media in mold with arbitrarily-shaped geometries. Thus, this non-conventional approach is now well understanding as applied to RTM process.

Printed in the United States
By Bookmasters